International Conference on Human Factors 2016

London, United Kingdom
28 - 29 September 2016

ISBN: 978-1-5108-8333-8

Printed from e-media with permission by:

Curran Associates, Inc.
57 Morehouse Lane
Red Hook, NY 12571

Some format issues inherent in the e-media version may also appear in this print version.

The Royal Institution of Naval Architects
8-9 Northumberland Street
London, WC2N 5DA
United Kingdom

Phone: 020 7235 4622
Fax: 020 7259 5912

publications@rina.org.uk

Additional copies of this publication are available from:

Curran Associates, Inc.
57 Morehouse Lane
Red Hook, NY 12571 USA
Phone: 845-758-0400
Fax: 845-758-2633
Email: curran@proceedings.com
Web: www.proceedings.com

CONTENTS

SAFETY BARRIER INTEGRITY IN SHIP OPERATIONS

S Mockler and **T King,** DNV GL, U.S.A.
T Relling, DNV GL, Norway

SUMMARY

Maritime safety regulations and inspection regimes have principally developed reactively in response to accidents and incidents in the industry. For much of the world's fleet the assurance of safety continues to be driven by demonstrating compliance with regulations and reacting to what we are able to measure, after the event.

In this paper, the authors present an overview of the *system approach* to safety and methodologies which may be used to apply this concept in the Maritime industry. This is based around two core principles: barrier management as a means to control occupational safety and major accident risks and a model for the assessment of safety culture in the Maritime industry to understand its impact on barrier integrity.

The paper concludes with a discussion of the potential value that this more holistic and dynamic approach to safety management may bring to the industry and what practical progress can be made to implement this unified model in ship design and operation.

1. THE STATUS QUO

1.1 WORLD FLEET ACCIDENT STATISTICS

The world fleet has exhibited a steadily downwards trend in the rate of maritime incidents (and particularly total loss rates) over the last 25 years. This contrasts with a significant peak in serious incidents during the shipping boom of the mid-to-late 2000s but which is now trending towards previous levels [1].

Despite this generally positive progress however, shipping remains a comparatively dangerous industry. In the period 2003-2012, shipping industry accidents accounted for an average of 900 crew and passenger fatalities per year. This is estimated to correspond to a crew fatality rate of six per 100 million work hours (10 times that of industry workers in OECD countries) [2].

1.2 OCCUPATIONAL VS. MAJOR ACCIDENTS

Developments in safety theory in recent years have given lie to the previously commonly held principle that good performance with regards to occupational safety (slips, trips, falls and work related injuries) provides an insight into vulnerability to major accidents or ship casualty events.

While an occupational accident (typically) involves a relatively simple chain of causal events, ship casualty events are (again, typically) substantially more complex, with multiple causes and inter-connected events. This makes it inherently more difficult to comprehend our vulnerability to them, since in addition to being outside of the personal experience of the typical operator, the connections and causal links between different events with the potential to lead to a major accident are not always clear.
We can find an illustration of the challenge inherent in this in recent, tragic history. On the 7th of May 2009, a rig owner and operator was awarded the US Government Mineral Management Service's Safety Award for Excellence (SAFE) on account of "outstanding drilling operations" and a "perfect performance period". This assessment was borne out in the figures, with one of the rigs in question showing such consistent safety performance that some 11 months after the award was received its safety record was being celebrated during a visit from a number of owner's and charterer's executives [3], when fire and explosions tore through the rig following a uncontrolled well blowout, with tragic consequences.

There are many complex lessons to be learned for major accident prevention and response from the Deepwater Horizon incident, but for the purposes of this paper the authors will focus on only one: that we must find a more holistic approach to assessing safety performance to address high risk activities involving complex technical and organisational systems.

1.3 THE HUMAN ELEMENT

The appreciation of the importance of the human element in maritime safety has grown steadily as we have applied advances from other industries in addition to our own hard-won lessons. From the adoption of the ISM Code onwards there has been an increasing focus from regulators and stakeholders on more operational safety measures and the role of the men and women in the maritime industry in maintaining and implementing these.

However, this increased focus and maturity of thinking remains at odds with our most visible response to incidents as an industry. Our first (and perhaps very human) response to an incident arising from a human error is still all too often to blame the operator, blame the watchkeeper or blame the Captain. While we now recognise the importance of human error to accidents at sea, we still tend towards a blame culture where it is viewed as the cause of an incident rather than a symptom of underlying issues within the organisation.

Even as the recognition of human and organisational factors has improved and grown, the actual effectiveness of our response must also be queried. A not atypical response to an incident caused by human error will take one of 2 forms:

- Write a new procedure, or *"Someone made an error so we should write a document that tells someone else how not to repeat this error."*
- Turn to technology, or *"Someone made an error so we must design a system that eliminates the need for a human decision."*

Both of these responses are perfectly valid and effective, when used proportionately and with due regard for their other consequences; significantly they are clearly measurable and show that the organisation is acting on an issue. Unfortunately, they are respectively contributing to the administrative snowballing of the maritime industry [4] and the growing unfamiliarity of seafarers with the complex, highly automated systems which they are expected to operate and maintain.

It is the opinion of the authors that we must as an industry alter our perspective on safety if we are to continue our positive trends and realise the further improvements in safety performance that wider society will expect of us. A critical part of this must be the further integration of our approach to the human element into other safety methodologies.

2. A SYSTEM APPROACH TO SHIP SAFETY

2.1 THE SHIP AS A SYSTEM

A ship, together with its crew and management may be described as a system, where technology, people and the organisation work together in a more or less coordinated fashion to operate the ship and achieve its business objectives. This is summarised graphically in Figure 1.

All aspects of the ship's performance are in some way influenced by the interaction of these three elements and safety performance is no exception. The most reliable assessments of an organisation's safety level are found through methodologies which address all three of these elements. While one or another may be dominant for a particular aspect of safety performance, methodologies that only address one of these elements carry the risk of blind spots which may not reveal all symptoms, causes or vulnerabilities. By triangulating findings from methodologies suited to all three of these elements, a far truer and more complete picture of safety can be obtained.

The same principle holds true for the management and improvement of safety performance, namely that the best results may only be achieved by treating all three elements of the safety system in appropriate measure.

2.2 A RECOMMENDED APPROACH

This paper briefly explores two models designed to analyse and evaluate ship safety at an organisational level, namely: barrier management and safety culture. In the context of the ship system, technological issues are extremely well addressed by barrier management, while safety culture assessment methodologies provide excellent insight into the human element and 'people' factors. Organisational factors are addressed in different ways by both methods.

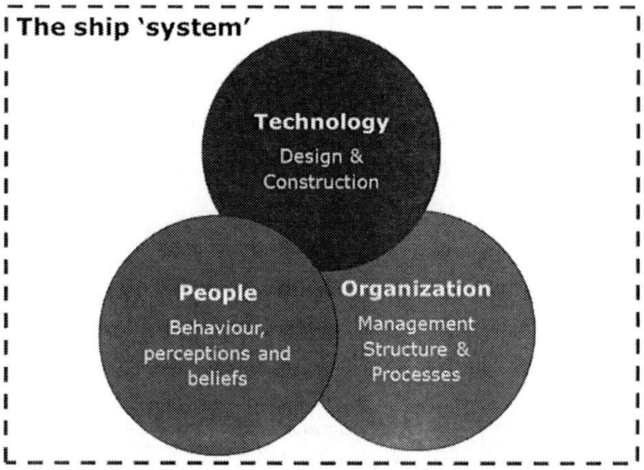

Figure 1: Elements making up the ship as a system

3. BARRIER MANAGEMENT

3.1 BARRIER THEORY

The concept of safety barrier management is perhaps most aptly summarised by Reason's 'Swiss Cheese' model (Figure 2), in which an organisation protects itself from hazards through 'defence in depth' provided by successive layers of protection (i.e. barriers) which either prevent the hazard from being realized and prevent or mitigate the associated risks. 'Holes' in the barriers provide a potential pathway for risks associated with a hazard to materialize as an incident.

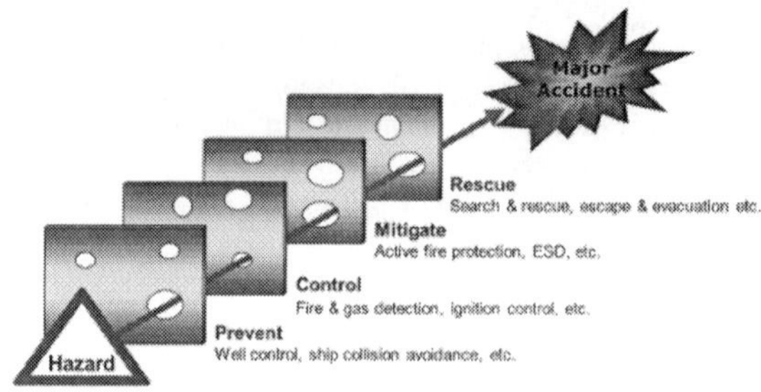

Figure 2: The 'swiss cheese' model, adapted from [5].

To understand which barriers exist or should exist, it is useful to apply a Bow-Tie approach. This method, shown in Figure 3, has its origins in fault tree and event tree failure analysis techniques. By defining a 'top event' with distinct causes and potential consequences, it is possible to identify preventative barriers which serve to reduce the probability of occurrence of the event and mitigating barriers which serve to reduce the consequence. The bow-tie is a powerful means of demonstrating the value of barrier-based methodologies for understanding the level of protection against complex, major accident or ship casualty events which involve the failure of multiple barriers.

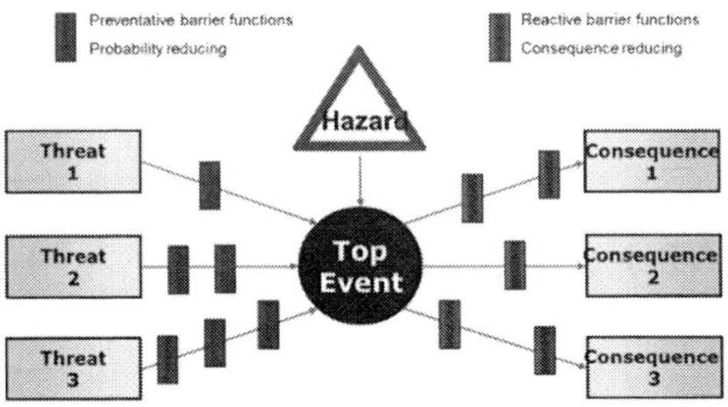

Figure 3: Generic Bow-Tie structure illustrating preventative and mitigating barriers.

Using this concept, barriers can be described at their highest level as having a *function*: a purpose or a role in the prevention or mitigation of an event. Barriers will only be able to fulfil (or realize) their function effectively if they retain their either integrity. Barrier integrity is reduced either from *active* failures of barrier function or *latent* degradation which weakens the barrier's effectiveness.

Barrier functions are sufficient to describe the overall layers of protection provided against a hazardous event and should, if successfully realized, have a direct and significant effect on the hazard and/or event sequence. Some examples are given in Table 1.

3.2 BARRIER ELEMENTS

Barrier functions alone do not however provide sufficient detail to understand the interaction between different components of the ship system. To do this, we define barrier *elements*: technical, operational or organizational measures which alone or together realize one or several barrier functions. The authors present a model which applies many of the conventions of [6] but adopts a more simplified approach.

Technical barrier elements are inherent in the design and construction of a ship and can consist of a wide range of systems, structures or other features which fulfil or contribute to barrier functions. Technical barrier elements may be either *active* (dependent on some form of positive action or input from an operator or control system); or *passive* (integral parts of the design of the ship which perform their functions without external influence). Some examples of technical barrier elements are given in Table 2, with this distinction highlighted.

Table 1: Examples of barrier functions

Barrier functions		Purpose
Preventative barrier functions	Maintain watertight integrity	Prevent capsizing caused by e.g. collision or grounding
	Prevent incorrect loading	Prevent capsizing or loss of hull strength caused by incorrect loading
Mitigating barrier functions	Detect and shutdown	Detect fire and shutdown ventilation system to mitigate consequences of fire
	Extinguish fire	Mitigate further consequences of fire through applying extinguishing medium

Table 2: Examples of technical barrier elements

Active elements	Portable firefighting appliances
	Watertight doors
	Quick closing valves
	Fixed fire extinguishing systems
Passive elements	Insulation of fire boundaries
	Deck drains
	Hull plating
	Escape signs and lights

The technical barrier elements identified above have been intentionally selected to align with well-established and understood safety measures required by existing maritime regulatory frameworks. This has the advantage of making the concept and structure more relatable for shipboard and shoreside personnel, but only addresses one element of the system model discussed earlier.

To effectively implement a barrier management structure, it is important that operational and organisational barrier elements are also included and identified.

Operational barrier elements typically consist of a specific task or procedure which needs to be performed (e.g. by an operator or a team of operators) to realize a barrier function. Operational barrier elements work either in concert with technical barrier elements to fulfil the barrier function or may be directly related to the activation of a technical barrier element.

An organisational barrier element by contrast, is related to the organisation's structure, processes and personnel responsibilities for these. Examples of each might be e.g. the dispatch of an assessment team on activation of a fire/smoke detector as an operational barrier, while the structure and training of the assessment team would be an organisational barrier element.

In practice, these two definitions are very closely related and it is not necessarily useful to distinguish between them, so the authors follow the convention of [1] and focus on operational barrier elements, while organisational barriers are generally treated as performance shaping factors to technical and operational barrier elements.

3.3 PERFORMANCE SHAPING FACTORS

A performance shaping factor may be defined as a human, workplace or organisational factor which has a significant effect on the performance of an operator or a team of operators with responsibility for a barrier. The SPAR-H model [7] defines eight performance shaping factors.

Available time
The amount of time that an operator or a crew has to diagnose and act upon an abnormal event.

Stress and stressors
The level of undesirable conditions and circumstances that impede an operator from easily completing a task.

Experience and training
The experience and training of operator(s) involved in a task.

Complexity
How difficult a task is to perform in a given context.

Ergonomics (including human-machine interface)
The equipment, displays and controls, layout, quality and quantity of information available from instrumentation.

Procedures
The existence and use of formal operating procedures for the tasks under consideration.

Fitness for duty
Whether or not the operator performing a task is physically and mentally fit to perform the task at the time.

Work processes
Aspects of working within the organization, including inter-organizational, work planning, communication, and management support and policies.

Four of these performance shaping factors are particularly relevant to the organisation (and to the second safety methodology) considered in this paper:

- Stress and stressors
- Experience and training
- Procedures
- Work processes

While this is not to say that the performance shaping factors not given above are not relevant to the organisation, available time and fitness for duty may be considered to be functions of work processes and procedures while complexity and ergonomics can be treated as functions of design.

By focusing on the performance shaping factors that are most closely tied to the organisation, we can begin to complete the picture of the ship as a system.

4. SAFETY CULTURE

4.1 DEFINITIONS AND THEORY

The concept of "Safety Culture" is one often discussed, frequently referred to and variously understood. Different academic and industry models exist [8], seeking to define and contextualise this rather broad and amorphous concept, however all share certain common aspects. Namely, that safety culture relates not only the behaviour of members of an organisation but also their perceptions or beliefs and that an organisation's culture is influenced both by individuals and its management structures and processes.

The author's approach to safety culture is based on Hale's model [9], and uses the following definition:

"Safety Culture is the elements or parts of organisational culture that influence the organisational members' attitudes, beliefs, perceptions, and behaviours, which have an impact on the level of safety within the organisation."

Drawing on Hale's definition, the authors use eight distinct 'dimensions' to measure safety culture.

Collaboration
The quality and amount of cooperation and support among colleagues, and between employees and managers, when it comes to working towards common health and safety goals.

Competence
An organisation's ability to establish and maintain employee knowledge and skills related to health and safety.

Managing conflicting goals
Employees' prioritization of health and safety when other goals or demands put their health and safety at risk.

Compliance
How relevant, available and understandable health and safety procedures are and the degree to which employees follow these procedures.

Incentives
The mechanisms that are in place for motivating employees to behave in a certain way related to health and safety.

Organisational learning
How an organisation uses feedback from reporting systems and other learning arenas to systematically reflect on ways to improve health and safety through employee attitude and behavioural change.

Creative worry
Employees' willingness and ability to imagine what could go wrong next and make suggestions for what should be done to reduce that risk.

Error tolerance
The extent to which an organisation is prepared against employee error, acknowledging that human error is inevitable and that they are often symptoms of weaknesses in the organisation.

4.2 ASSESSMENT METHODOLOGY

The authors apply a mixed methods approach [10] to the assessment of safety culture, combining quantitative and qualitative methods to improve the accuracy of results. Quantitative methods using surveys or questionnaires to gather data provide a strong basis for comparison and benchmarking as well as allowing the coverage of a large number of respondents. Qualitative methods using interviews, focus groups, direct observations etc. allow much richer and more descriptive information to be gathered and provide greater opportunities for insight into underlying causes or themes and patterns.

Quantitative analysis of survey responses can highlight areas for detailed investigation, both through comparison to benchmark results (example in Figure 4) and through the identification of significant differences between results for different parts of the organisation. Testing for significant differences and the influence of independent variables (such as operating fleet, rank, department, age or nationality) can give valuable indications of sub-cultures and influencing drivers.

The initial quantitative analysis may be equated to a maturity level of safety culture in each of the eight dimensions. A hierarchy of maturity levels is presented in Figure 5, based on [11].

Ignorant
Fear of non-compliances being discovered is the only driver for safety performance.

Ad-hoc
Safety is recognised as important in its own right and the organisation reacts to accidents and incidents.

Bureaucratic
The organisation has mapped and manages hazards but depends heavily on the associated management system.

Aware
Safety leadership and values drive continuous improvement in safety behaviour.

Learning
Safety is a core business value and is an integrated part of operations and strategy.

Figure 4: Example of benchmarking analysis

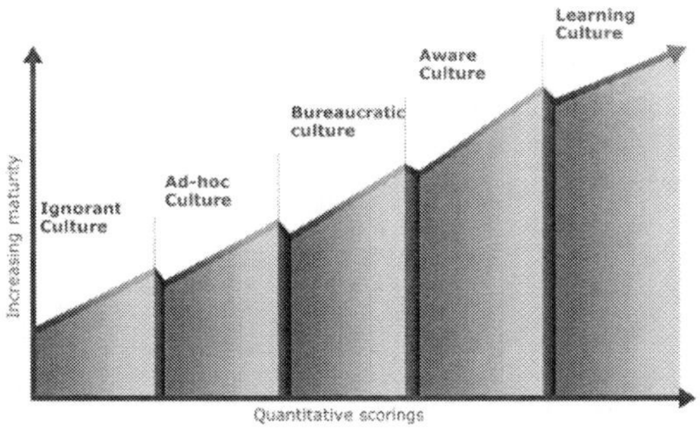

Figure 5: Maturity levels of safety culture in ascending order.

While quantitative techniques can give an effective measure of overall maturity, these should be properly validated against findings from qualitative methods to improve confidence in the results.

4.3 FROM ASSESSMENT TO IMPROVEMENT

While there is inherent value to an organisation in gaining an increased appreciation of its safety culture, the greatest benefits are only utilised by leveraging this to drive and implement safety initiatives targeted at improving the safety culture. These measures will be highly specific to both the organisation in question and the nature of the findings of such an assessment.

If significant differences are identified between groups within the organisation, against benchmarks or indeed against the organisation's desired state, then a more detailed analysis of the related survey responses can indicate potential improvement areas. By correlating these with root cause findings from the qualitative methods, very specific safety improvement areas can be identified and appropriate measures implemented.

5. TOWARDS A UNIFIED MODEL

5.1 PRACTICAL BARRIER MANAGEMENT

DNV GL is currently piloting a barrier management approach based on the existing scope of Classification surveys that seeks to build on the current maritime safety frameworks, while adapting them into a new context which provides a more secure overview of safety resilience. This approach makes the relative importance of individual safety measures far clearer in the context of main barrier functions. It also provides a powerful means of presenting and trending the level of safety for the ship at different times of assessment.

A barrier structure consisting of high level barrier functions is developed for specific major accident risks, as illustrated in Figure 6 for machinery space fire risks.

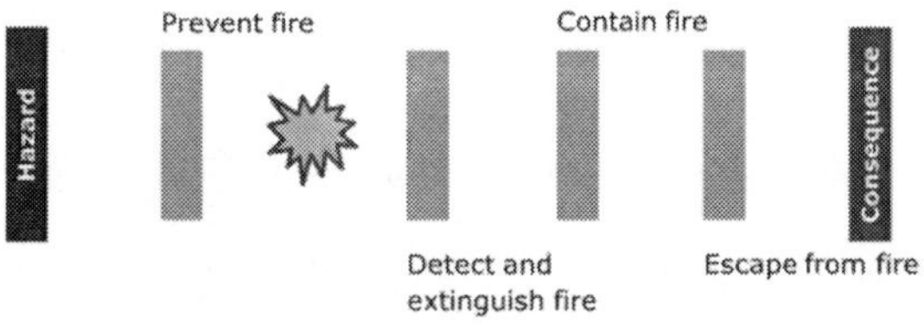

Figure 6: Barrier functions for machinery space fire risk.

Individual safety measures assessed under the Classification survey regime are then grouped under the relevant barrier function as barrier elements, as illustrated in Figure 7.

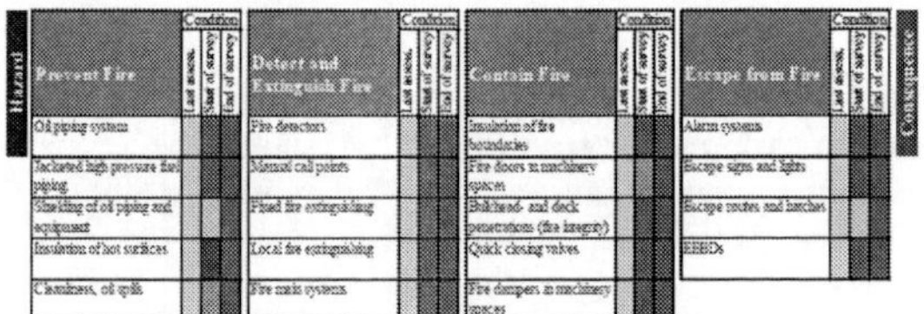

Figure 7: Barrier elements for machinery space fire risk.

Once a barrier element structure is established, each element can be assigned integrity criteria. In the context of technical barriers, these criteria are based on the reliability and availability of the barrier element in realizing the barrier function. In DNV GL's approach, these criteria are initially based on classification and statutory requirements and are then expanded upon according to the ship operator's own standards and operational safety criteria.

Findings from DNV GL Classification and delegated statutory surveys are then assigned to the appropriate barrier elements and assessed against the relevant integrity criteria to provide an integrity rating.

5.2 SHAPING BARRIER PERFORMANCE

As discussed in section 3.3, four performance shaping factors are considered to be particularly connected to organisational safety factors and the safety culture methodology outlined in section 4. In particular, the authors propose that direct correlations can be made between performance shaping factors and safety culture dimensions in some cases, identified in Table 3.

Table 3: Correlation between performance shaping factors and safety culture dimensions

Performance Shaping Factor	**Safety Culture Dimension**
Stress and Stressors*	Managing conflicting goals
Experience and Training	Competence
Procedures	Compliance
Work processes	Collaboration

* While there is a strong correlation with this performance shaping factor, other organisational and external factors remain important.

The barrier management methodology outlined by the authors in section 5.1 does not yet include a quantitative link between barrier elements and the associated performance shaping factors; however performance shaping factor data can and should be considered when defining barrier integrity criteria to reflect unique organisational aspects. Further work to model the influence of performance shaping factors on individual safety barrier elements is planned.

6. CONCLUSIONS

The authors have presented a case for approaching the ship as a system, influenced by interacting technical, people and organisational factors. An overview has been presented of two particular methodologies for the assessment and management of safety in the maritime industry, each addressing different aspects of the ship safety model.

Through the increased application of barrier management methodologies, the authors believe that the maritime industry can substantially improve its resilience to major accidents. A well-structured barrier methodology can provide significant insight into the degree of 'defence in depth' which exists for a given ship or organisation. If coupled with well-defined integrity criteria, barrier-based techniques can also improve the effectiveness and interpretation of existing inspection and survey methods by integrating them into a more dynamic barrier management strategy.

To supplement the technical and operational strengths of barrier management techniques, a model for safety culture assessment was presented and discussed. Safety culture assessment provides an efficient means of measuring human factors in ship safety and of assessing broader aspects of the organisation with an influence on safety. The definition of distinct dimensions of safety culture, combined with a mixed methods approach to assessment allows a reliable quantification of safety culture maturity in a manner which is meaningful to ship operating organisations.

Finally, an example of a practical methodology for the application of barrier management within existing maritime safety frameworks is presented. The methodology is presently being trialled and has so far proved to be a powerful means of presenting barrier integrity at both an individual ship level and for a fleet as a whole.

The authors hold the opinion that the integration of safety culture tools into barrier management methodologies has significant value as a way to further our understanding of the ship as a system and ensure that we can develop a holistic approach to safety. Further work is planned to explore ways in which this integration can be achieved, however an initial direction is identified through the correlation of safety culture dimensions with specific performance shaping factors.

7. REFERENCES

1. ASTRUP, O.C. et al, 'A Framework Addressing Major Accident Risk in the Maritime Industry', *SNAME World Maritime Technology Conference*, 2015.

2. LONGVA, T. et al, 'The Future of Shipping', *DNV GL, Høvik*, 2014.

3. UNITED STATES COAST GUARD, 'Report of Investigation into the Circumstances Surrounding the Explosion, Fire, Sinking and Loss of Eleven Crew Members Aboard the Mobile Offshore Drilling Unit Deepwater Horizon, Volume I', 2010.

4. DANISH MARITIME AUTHORITY, 'Administrative Burdens in the Maritime Sector - Summary Report', 2013.

5. REASON, J., 'Managing the risks of organizational accidents', *Ashgate, Aldershot*, 1997.

6. FLØTAKER, H. et al, 'Good Practices - Barrier management in operation for rig industry', *DNV GL Technical report 2013-1622*, 2014.

7. GERTMAN, D. et al, 'The SPAR-H Human Reliability Analysis Method', *Technical Report NUREG/CR-6883, U.S. Nuclear Regulatory Commission Office of Nuclear Regulatory Research, Washington DC*, 2005.

8. HEALTH & SAFETY EXECUTIVE, 'A review of safety culture and safety climate literature for the development of the safety culture inspection toolkit', *HSE Research Report 367*, 2005.

9. HALE, A., 'Culture's confusion', *Editorial in Safety Science 34, 1-14*, 2000.

10. LISTYOWARDOJO, T.A. et al, 'Mixed Methods: Improving the assessment of safety culture in healthcare', *DNV GL Strategic Research & Innovation Position Paper 11-2014*, 2014.

11. PARKER, D. et al, 'A framework for understanding the development of organizational safety culture', *Safety Science 44 551-562*, 2006.

8. AUTHORS BIOGRAPHY

Simon Mockler holds the current position of Senior Risk Engineer at DNV GL, based in Miami. He is responsible for the delivery of safety and risk management services in the Americas. His previous experience includes safety culture assessment, safety management system development and technical risk assessments in the Maritime, Oil & Gas and Offshore Wind industries.

Tobias King holds the current position of Special Projects Engineer at DNV GL, based in Miami. He is responsible for the development of barrier-based safety methodologies with additional approval and survey responsibilities for ships in operation. His previous experience includes Class rule development and safety aspects of alternative fuels in shipping.

Tore Relling holds the current position of Senior Consultant at DNV GL, based in Ålesund. He is responsible for the development and delivery of DNV GL's human and organisational factors advisory services in the Maritime industry. He has been responsible for a large number of safety culture, marine risk management and human performance projects since joining DNV GL. His previous experience is in Aviation and Leadership Development in both the military and commercial sectors.

Human Factors, 28-29 September 2016, London, UK

THE HUMAN ELEMENT OF MODULAR CAPABILITY

C Kelly, K Sharp Ltd., UK, **G Fletcher,** Frazer-Nash Consultancy Ltd., UK, **R Crabbe,** BMT Defence Services Ltd., UK, **M Wood**, BMT Reliability Consultants Ltd., and **D Collins**, University of Central Lancashire, UK.

SUMMARY

This paper discusses the Human Factors issues surrounding naval ships designed with modular capabilities. Modularity is not a new concept, but the training and personnel implications of modularity on a ship's core company are not well understood. With vessels such as the Type 26 frigate, the adverse impact on the core crew, and the infrastructure that would be needed to maintain their training and experience, could be significant. The proposed 'UK Way Ahead on Modularity' highlights an equipment-based description that, although useful from an industrial procurement perspective, does not adequately articulate the whole concept of modularity. In particular, changing naval ships from being 'flexible' (ready to perform all roles) to 'adaptable' (only performs certain roles after a change or addition in modular crew and systems) could cause substantial effects at a fleet capability and political level. The study developed a framework to understand both the equipment and Human Factors issues.

NOMENCLATURE

CT	Collective Training
DHCSTC	Defence Human Capability Science & Technology Centre
DLOD	Defence Lines of Development
Dstl	Defence Science & Technology Laboratory
FOST	Flag Officer Sea Training
FTR	Fit to Receive
HC	Human Capability
HF	Human Factors
LCS	Littoral Combat Ship
MHC	Mine countermeasures and Hydrographic Capability
MOD	Ministry of Defence
NCHQ	Naval Command Headquarters
OCV	Offshore Combatant Vessel
PLOD	Personnel Line of Development
RN	Royal Navy
T26	Type 26
TLOD	Training Line of Development
UK	United Kingdom
US	United States

1. INTRODUCTION

1.1 BACKGROUND

The decreasing hull numbers in naval fleets and the increasing diversity of operations require greater capability within each vessel. With the longer service life for modern vessels, relative to the technology and equipment embarked within them, flexibility is gaining a greater importance within a design. A method of achieving this without increasing crew and limiting the hull principal dimensions is through role reconfiguration and "modularity". Whilst this may have a large impact on rank structures, complement composition and training, it is a model more and more navies are turning towards [1] [2].

The UK Defence Human Capability Science & Technology Centre (DHCSTC) initiated a research project to provide evidence to inform development of the policy, processes and practices required to ensure effective management of the human component of modular capability.

In particular, evidence was required by Naval Command Headquarters (NCHQ) to inform the development of optimised solutions for Capability modules of the T26 Global Combat Ship and MHC programmes, including the drawing of lessons from modular capabilities employed today (within and out with Defence).

This paper presents an overview of the study that was carried out [3].

1.2 STUDY AIMS AND SCOPE

The overall aim of the study was to better understand the pan-DLOD implications for operational capability of platform modularity on Human Capability (HC) in general and specifically the Training and Personnel Lines of Development (TLOD and PLOD). The objectives of the initial phase of the study (reported here) were:

- To identify and document the breadth of HC and training-related questions that should be considered in relation to platform modularity;
- To develop a definition of platform modularity;
- To develop a modularity framework, which might include different levels of capability, to establish a context for deeper investigation.

The study did not aim to provide the measures required to implement a training and manpower regimen to underpin modular equipment concepts, rather it sought to identify the areas in which further research is needed and likely to be of benefit to the execution of modular capability. Effectively, it was a scoping study to inform directions for decision making and future research.

1.3 METHODOLOGY

First, the study team conducted a literature review on modularity, in particular to draw any lessons from contemporary or previous examples of modular capability within Defence. The review also briefly addressed the topic of adaptive expertise, which was of interest to NCHQ. Second, a series of interviews were conducted with relevant military stakeholders engaged within the training and personnel domains and also with personnel who deliver modular style capabilities at present. Together with the literature review and several team workshops, an overarching definition of modularity was then established to include the human component as well as equipment. Third, a 'modularity framework' was developed that would enable an initial level of modular system classification. Lastly, drawing on all of the documentary evidence, a series of questions was derived for deeper analysis of the HC issues in a further research study.

In order to provide a repository for the information gathered during the study, all relevant documentary information together with input from stakeholders and workshops was collated in an Excel workbook.

1.4 ASSUMPTIONS

At the outset, a number of assumptions were made:

- The 'core platform' in a modular capability is essentially able to Float and Move although a limited self-protection (i.e. Defend) capability is assumed to be in place (upper-deck weapon crews, limited operations room manning etc.). However, in order to be fully operational, the core platform will need to embark operational capability modules.
- Three levels of ship training have been used to describe levels of overall ship capability or the level of assurance to which a ship can be trained: Safe at Sea; Safe to Operate; and Mission Ready.
- Fit to Receive (FTR) – is the term developed to define the required state of readiness to which parts of the core platform and modular equipment in order to be activated or to accept additional/different equipments and/or people; this is more than just maintaining the interfaces for equipment.

2. LITERATURE REVIEW

2.1 UNDERSTANDING MODULARITY

Modularity is not a new concept and its origins go back to at least the 1970s with, for example, the United States (US) Navy's Sea Systems Modification and Modernization by Modularity programme [4]. The term modularity has been defined in numerous ways over the years, but simply put it means the creation of interchangeable components or parts of a complex system linked together to perform desired tasks or missions through a set of common standards and interfaces.

For the UK, modularity is also part of a wider approach to ship 'adaptability', which can be thought of as an intermediary stage between 'flexibility' and 'conversion' in the spectrum of a platform's ability to meet change. At one end of the spectrum, and built in from the start, is the platform's flexibility to meet the full range of possible operational mission requirements for the type; at the other end of the spectrum is the platform's conversion, where it is completely role changed (e.g. the conversion of a supply vessel to a helicopter training vessel). In between, is the platform's capacity to be upgraded or updated through-life [5].

Many navies across the world have developed modularity programmes. The review examined seven countries in addition to the UK: Denmark, Germany, France, Canada, Australia and US. The Royal Danish Navy is widely credited with pioneering work in warship modularity, with its Standard Flex (STANFLEX) concept. The STANFLEX system uses a set of rapidly exchangeable, mission-oriented containers deployed into common spaces. Among the mission-oriented containers available are those for Anti-Air Warfare, surface warfare, mine countermeasures, Anti-Submarine Warfare, oceanographic research and pollution abatement. Importantly, the configuration of the modules is decided in the preparation phase of a deployment, and the configuration stays the same until the ship returns to home port. The Royal Danish Navy does not change modules in overseas operation areas.

The French Frégate Européenne Multi-Missions (FREMM) programme is designed to restore the operational capabilities of the Navy's frigates and destroyers. A total of 17 ships are planned to be built, the first eight frigates in the series having being ordered in November 2005. The FREMM is intended to have a "lean manning" of 108, which is half the number required for frigates of the previous generation.

The Royal Canadian Navy has some experience of modularity though far less than other navies such as the Danish or US. A Defence Research and Development Canada review of 'modular capabilities' concluded that they would provide the Navy with potential benefits in terms of mission flexibility, upgradeability and overall cost, but cautioned that modular capabilities should "not be a key driver of the overall ship architecture" [6].

The Australian 2012 Defence Capability Plan indicated the potential for Australia to rationalise its patrol boat, mine countermeasures, hydrographic and oceanographic capabilities into a single modular multi-role class or family of around twenty Offshore Combatant Vessels (OCV). The Australian government requires the OCV to undertake additional roles to that of the current Patrol, Hydrographic and Mine Counter-Measures (MCM)

Force, namely Maritime (Littoral) Warfare, counter-terrorism, counter-piracy, Special Forces Support, etc. Central to the capability provided by the OCV will be the ability to readily adapt and enhance baseline sea frames with modular 'capability brick' combinations of systems, both materiel and support, with appropriately trained personnel embarked to undertake one or more primary roles during periods of operational availability. Role reconfiguration of OCVs is envisaged primarily as a deliberate, pre-planned activity at dedicated support bases [7].

In the US, the Littoral Combat System (LCS) is a surface combatant equipped with modular "plug-and-fight" mission packages for countering mines, small boats, and diesel-electric submarines, particularly in littoral (i.e., near-shore) waters. Two very different LCS designs have been built (see Figure 1). The LCS employs automation to achieve a reduced-sized core or sea-frame crew. The programme's aim was to achieve a core crew of 40 sailors although the Navy has since increased that number to about 50 in order to better address workload and watch-standing requirements.

Figure 1: US Littoral Combat Ships

As has been widely reported, the LCS program has been heavily criticised due to budget overspends, design and construction issues including the ship's ability to withstand battle damage, concerns over whether the ship is sufficiently armed, its "lean crew", and its overall operational effectiveness [8] [9] [10] [11].

For example, amongst many criticisms made, the US Government Accountability Office (GAO) concluded: "core crews depended heavily on sailors from the surface warfare mission-module crew to perform watch standing, training drills, and engineering maintenance over the course of the deployment. This is a departure from what was envisioned in the concept of operations, and creates a situation that may not be sustainable under a different pace of operations or with a different mission module deployed" [10].

2.2 ADAPTIVE EXPERTISE

Adaptive expertise may best be thought of as the capability, commitment and confidence to think in an agile or adaptive fashion and to action such thoughts. In contrast to routine expertise, individuals with adaptive expertise possess the knowledge of why and under which conditions certain methods have to be used or new methods have to be devised [12].

The construct of adaptive expertise would appear to offer the following advantages:

- Increased flexibility in core crew members, offering the potential to decrease the number of specialist staff which must be assigned as role-purpose changes;
- A reduction in the problems of team cohesion and execution as rivalries are mitigated by the combination of flatter structures, increased cross specialism recognition and respect and shared expertise;
- Potential reduction in skill fade: the 'deeper' knowledge and training inherent in the application of adaptive expertise has been shown to be associated with better retention of skills together with the ability to transfer knowledge into related or novel settings;
- Enhanced performance under stress, whether this pressure be caused by time available, fatigue, combat challenges, or a combination thereof.

At the very simplest level, adaptive expertise results from environments in which learners are stimulated to explore and/or discover. However, although adaptive expertise may be developed by current methods of training, optimal return may require some changes to the current ethos and approaches used. It would seem that the main challenges to the introduction of training for adaptive expertise may relate to time needed and the cultural evolution required of trainers and learners.

2.3 HUMAN CAPABILITY ISSUES

The literature review revealed a wide range of HC issues. A total of over twenty HC 'themes' were identified, from adaptability to watchkeeping. Each of these themes was broken down into a number of sub-themes or topics. For example, under the theme of manning, the topics were: churn; 'hybrid sailor'; lean manning; optimised manning; reduced manning; safe manning; scalable / 'flex' crew; and Squad system.

The size and complexity of the manning (or complementing) task for new naval vessels is significant. Even small errors can have major cost implications; if the complement is too small, this may compromise capability and possibly require extra equipment to be installed to compensate for the shortfall; if too large a complement will increase the Whole Life Costs of the necessary accommodation in the platform. The risks of reduced manning appear to be well acknowledged in the literature, the experience of the US LCS being a prime example [10]. Review of the T26 core complement identified a number of shortfalls as a result of adopting the solution of 118 personnel, i.e. "limited capacity to concurrently recover any damaged systems whilst retaining the ability to fight." [13].

The information gathered under these topics, was analysed and synthesised to derive a number of questions for further research.

3. STAKEHOLDER CONSULTATIONS

3.1 INTERVIEWS

The objective of the interviews was to establish the likely HC issues and concerns, e.g. manning and training for future Royal Navy (RN) modular capabilities, and also to identify any lessons from existing modular approaches. Candidate organisations focused on: Force Generating Authorities, training policy organisations, and organisations currently supplying modular-type capabilities to ships.

A total of nine military stakeholder organisations were interviewed. Additionally, one non-military organisation – a maritime and sub-sea construction company working in the oil and gas sector – was consulted.

The questions for the interviews were developed following an initial review of the literature and a number of workshops and meetings between the study team members regarding the key issues for modularity. Example questions used as themes for discussion in the interviews included:

- How do you think platform modularity might/could work?
- What level do/could you have as a 'module'?
- How will personnel also be tasked in a modular manner?
- How will a 'stock' of competent personnel for each capability be maintained?
- How will these or competent personnel be managed when away from the platform?
- How will 'modular personnel' be integrated swiftly and fully with native platform personnel to ensure delivery of the whole capability?
- What do you think will be the enablers and/or barriers to implementation of a modular capability approach?

3.2 WORKSHOPS

In order to consolidate findings and information gained through the literature review and stakeholder interviews, four workshops were conducted. The information collated in both the literature review and the stakeholder interviews was analysed and distilled during the workshops to characterise not only the framework and definition, but also to establish the further research questions.

The workshops allowed the multi-disciplinary study team to examine the issues raised and propose a set of research questions that range from the tactical level (relevant to a Project/Programme team aiming to deliver a modular set of capabilities) to the strategic level (aimed at understanding how a modular capability should be managed within the RN/MOD/Defence Enterprise).

Lastly, the output from the workshops (in terms of characteristics of modular systems and definitions) were discussed again with certain stakeholders to confirm and, as far as possible, to validate the findings.

3.3 KEY FINDINGS

A number of common themes and questions emerged from the military stakeholder interviews, although different organisations inevitably had slightly different perspectives. Key findings included:

- Modular units need a shore base to provide storage and a 'home' for personnel with meaningful employment opportunities, training and career management. Being able to maintain a suitable level of skill when *not* embarked is essential.
- The ship needs to maintain a FTR capability to ensure modules can be effectively integrated. Collective Training (CT) is seen as critical to this, but in some situations when this is not possible, proxy teams can help the ships develop the processes and ways of working.
- Good interpersonal relationships between core and modular personnel are important; training or past-history support this. Developing trust between teams is essential for safety critical capabilities.
- Existing modular-type capabilities are supported by strong infrastructure and force generating processes. Challenges exist in how to assure the newly formed capability following a modular integration, particularly if it takes place after FOST work-up.
- There is some concern about how maintenance and engineering roles will be supported in a modular platform.
- Pan-DLOD integration and attention to Concepts and Doctrine are critical when developing modular capability.
- Operating in a modular manner requires an adaptable mindset amongst personnel and in units, and for many areas achieving this will require a change in culture.
- Effective change management will be needed to introduce changes around modular capability.

In contrast, in the oil and gas domain, modularity does not translate into people following equipment around; instead with the addition of a different module, the crew is retrained on the new equipment. Furthermore, the most efficient use of the personnel has been found from matching crew members to each other and to the ship such that they develop good team cohesion and understand the eccentricities of their vessel.

4. FRAMEWORK DEVELOPMENT

4.1 MODULARITY DEFINITION

Drawing on the outputs of the literature review and stakeholder consultations, the following definition of modularity was derived:

"Modularity is the tailored enhancement of a basic core platform by the timely addition or changeover of subordinate human and/or equipment capabilities integrated to perform specified tasks or missions. This will require a set of common standards, practices and interfaces."

4.2 MODULARITY FRAMEWORK

To develop the modularity framework, the 'complexity' of the modular integration process was used as a discriminator to delineate the different levels of modularity. A system that has significant physical or human 'touch-points' in the core platform together with a high integration training demand and/or organisation change, is classified as a high complex modular system (e.g. a helicopter flight); a system which simply requires fitting and has limited interaction with the core platform and/or other modules is a low complex modular system (e.g. coalition communications system).

The various scales and types of modular capability, either as equipment or human elements together with a broad evaluation of what functions each element performs, are captured in Figure 2. In order to further develop the framework, the different modular classes and types from the Proposed UK Way Ahead on Modularity [14] were mapped to the modular categories above in order to establish a common theme between Equipment LOD and Training/Personnel LODs. Extrapolating from Figure 2 and the characteristics of the existing modular capabilities [14], a broad assessment was made about the level of shore infrastructure and support that would be required for each scale of modular system. The resulting framework (see Table 1) should assist decision makers in identifying the high level requirements for modular systems in relation to their perceived effect upon HC.

Table 1: Potential Consequences of Modular Capability Integration (extract only).

	Core platform	Equipment only module	Low Equipment, High Personnel Complexity Module
Impact on crew workload (0-10)	3-4	5-7	6-8
Modular Component Contin-uation Training Ashore	N/A	Likely to be limited to short training courses for operators and maintainers	Need to be a centre for training to maintain a level of proficiency but this is unlikely to fill 100% of the time available to the relatively low complexity nature of the task.
Manage-ment when disembarked	N/A	No	Modular capability teams will need career management as well as line management. A structure will need to be provided to give them scope for different employment and promotion.
Employment when disembarked	N/A	No	As training is unlikely to fulfil their full, disembarked, time, they should be employable in their professional role.
Shore infra-structure	Yes (Support) The core platform will need periods of extended maintenance requiring external support.	Limited; Storage infra-structure.	This level of training, employment and management will require a shore-based infrastructure to provide a location and a common focus.

4.3 REVISED DEPLOYMENT CONCEPT

The current Force Generation process for deploying ships aims to bring all elements together for CT prior to deployment. Individual training takes place prior to joining the ships and then there is a period of collective training in the ship prior to deployment to the intended theatre of operations. In the training package that the ship undertakes, there will be elements to include core warfighting (and associated damage control etc.) and also mission specific elements such as counter-piracy boarding. In addition, the ship will receive training on contingent tasks that they may have to undertake during their deployment such as disaster relief and non-combatant evacuation operations.

In the future, if the full utilisation and rapid re-deployment of modular capabilities is to be realised, there would need to be a cultural change in the way in which both ships and subordinate capabilities are generated for deployment and then kept in reserve until required. In other words, it will require a means by which capabilities can be rapidly integrated in theatre, both in equipment and people terms, when required by the operational commander. There will be varying needs for validation, assurance and integration depending upon the time, location and operational urgency in embarking a module. Clearly, the actual approach taken to modularising deployments will be dependent on high level doctrine and strategy and on the specific operational requirements of a mission.

5. PROPOSESD RESEARCH ACTIVITIES

Based upon the stakeholder interviews, literature review, modularity definition and framework, a set of further research questions were developed, which were aimed at further analysis of the issues that had been identified through the course of the study. These questions included:

- What is the validation process for the FTR component on the core platform, for the embarking modules, and for the overall new Capability generated? This encompasses the breadth of HC and training in addition to the extant RN equipment FTR processes.
- What are the implications of a modular capability approach on the manning structure of the core complement?
- What are the implications of modularity on the types and/or balance of training conducted at sea and at shore?
- What changes to CT might need to be made to support a modular capability approach, including modular redeployment?
- What are the implications of a module's physical complexity (low/high) and its integration complexity (low/high) on the capacity of the core crew to perform their jobs?
- What are the implications (benefits and barriers) of a possible break away from the 'Cap Tally' culture to a 'Trade' culture?
- What are the consequences of a modularity capability approach for personnel jobs / career management, particularly with respect to 'specialists' versus 'generalists'?
- How can it be ensured that crew fatigue problems (short-term and cumulative) are not worsened by the introduction of modular capability approach?
- How can it be ensured that crew fatigue problems (short-term and cumulative) are not worsened by the introduction of modular capability approach?
- What morale challenges could arise due to the unorthodox working arrangements associated with modularity?

For each of these questions, a number of subsidiary questions were derived. A rationale for each main question was also produced. For example, a strong factor in achieving current RN effectiveness is believed to be that the ship's component capabilities come together to train and rehearse long before deployment. During this work up, the crew develops knowledge of the ship, its systems and their roles, procedures and ways of working, team relationships, and a sense of ship's 'identity'. Effectively the crew learns to operate and inter-operate on both technical and social levels. However, true modularity will affect this traditional ship generation process. Therefore, it is essential that all the interfaces associated with a modular capability are developed ahead of time and maintained at a suitable level of readiness fit for when the ship receives it new module. There may be opportunities for further CT, but the main aspects of work-up need to be pre-prepared. This could be anything from maintaining the ship's physical equipment for linking with a module to understanding different roles and responsibilities for when additional crew come on board.

Research is therefore needed to understand the social and technical requirements for ship and modules to achieve FTR and to be received, readiness, and how this can be optimised in the worst case scenario, e.g. urgently when deployed at sea. What new technologies and manning approaches could assist this, e.g. distribute team training or remote maintenance support? Additional questions need answering around how to assure 'mission ready' with non-traditional work-up routes.

6. CONCLUSIONS

Modularity is not a new concept and has had many definitions and applications over the last 40 years. Many of the RN's existing capabilities have characteristics of a modular system with positive lessons that can be applied to a 'wholeship' modular capability concept.

Many of the inherent features of modularity are already in place, but conceptual, cultural and doctrinal changes may need to be addressed if modular concepts of employment are to be successfully implemented. Modularity may also necessitate a different approach to CT, particularly in modular redeployment.

A common theme throughout all the modularity projects analysed was the drive to reduce manpower in order to realise efficiency savings. However, modularity has potential implications on the core complement manning structure and employment. It might change the balance and types of training conducted at shore establishment and at sea, and the technical support infrastructure.

In order for a lean manned platform to be able to perform under all the demands expected of it, the adaptability of its crew may need to be enhanced. The development of an adaptive expertise skills base would be desirable for the RN in general and in particular to address modularity applications.

Irrespective of modularity, the adverse effects of lean manning on crew fatigue (short-term and cumulative) are continually under-estimated. Morale can be a challenge due to unorthodox and inflexible working arrangements.

Implementation of modularity reinforces the importance of a systematic, holistic approach to Capability Management. For modularity to work, ships must be "fit to receive", as well the module and associated staff being "fit to be integrated".

The proposed UK Way Ahead on Modularity highlights an equipment-based description that, although useful from an industrial procurement perspective, does not adequately articulate the whole concept of modularity. The move from a traditional flexible 'just in case' mind-set to the adaptable 'just in time' mind-set as created by modularity, will requires a shift in RN doctrine.

The study raised many questions about the implications for operational capability of platform modularity on Human Capability. Research needs to be conducted to answer these questions before the concept of modularity is taken forward.

7. ACKNOWLEDGEMENTS

The study was funded by the UK MOD through the DHCSTC Research Programme under Dstl contract DSTLX-1000069524 and from whom permission to publish was granted. Copyright: MOD, BAE Systems and contractual Suppliers.

The authors wish to express their thanks to all of the stakeholders who participated in the study.

8. REFERENCES

1. Kimber, A. and Giles, W., *'Minor Warship Roles - How technology is leading to a new vessel type'*, Paper presented at Pacific 2008, Sydney, Australia, 2008.

2. Jones, A. and Kimber, A., 'Balancing Survivability, Operability and Cost for a Corvette Design', Paper presented at *The Affordable Warship*, RINA, 20-21 June, Bath, 2012.

3. Crabbe, R., Kelly, C., Fletcher, G., Collins, D. and Wood, M., *'Technical Report 1 (Initial Study) TIN 2.044 Training for Platform Modularity'*, Defence Human Capability Science & Technology Centre, O-DHCSTC_I2_T_T2_044/002, 2015.

4. Lawson, C.E., *'SEAMOD - A new way to design, construct, modernize and convert U.S. Navy combatant ships'*, Advanced Technology Systems Division, Naval Sea Systems Command, 14th Annual Technical Symposium. Association of Scientists and Engineers (ASE), 1977.

5. Naval Command Headquarters, *Maritime Synthetic Training Strategy*, NCHQ, April 2014.

6. MacKenzie, S.C. and Tuteja, R., *'Modular Capabilities for the Canadian Navy's Single Class Surface Combatant. A Perspective on Flexibility'*, Ottawa: Defence Research and Development Canada, February 2006

7. Navy Strategic Command, *'Project SEA 1180. Patrol Boat, Mine Hunter Coastal and Hydrographic Ship Replacement. Offshore Combatant Vessel (OCV) Initial Capability Description'*, Australia NSC, Version 1.0. July 2011.

8. US Navy (2014) Small Surface Combatant Task Force. US Department of the Navy, Chief of Naval Operations. Memo from SECDEF to SECNAV and CNO, LCS Program.

9. McCain, J., *'Remarks by Senator John McCain on the navy's troubled Littoral Combat Ship program'*. Floor Statement, 9th April 2014.

10. Government Accountability Office, *'Littoral Combat Ship Deployment of USS Freedom Revealed Risks in Implementing Operational Concepts and Uncertain Costs'*, GAO-14-447, July 2014.

11. Government Accountability Office, *'Littoral Combat Ship Additional Testing and Improved Weight Management Needed Prior to Further Investments'*, GAO -14-749, July 2014.

12. Bohle Carbonell, K., Stalmeijer, R.E., Könings, K.D., Segers, M. and van Merriënboer, J.J.G., *'How experts deal with novel situations: A review of adaptive expertise'*, Educational Research Review 12, 14–29, 2014.

13. Naval Command Headquarters, *'Background notes on T26 GCS Core complement proposals'*. NCHQ, 19 March, 2014.

14. Courts, M.D., *'Proposed UK Way Ahead On Modularity'*, Defence Science and Technology Laboratory, Naval Systems, Dstl/TR75915 Issue 1, 10 January 2014.

9. AUTHORS BIOGRAPHY

Christopher Kelly is an HF Consultant at K Sharp Ltd. He has over 30 years' HF experience on a wide range of military and civil projects. Between 2008-12, he was actively involved in the MOD's HFIDTC research programme. More recently, he has worked on several DHCSTC studies. At the time of this research, he was employed at Symbiotics Ltd.

Dr Georgina Fletcher is a Principal Consultant at Frazer-Nash Consultancy. She has a background in psychology and ergonomics and over 20 years' experience working in research and consultancy across a range of organisations and domains, including defence, aviation, healthcare and the offshore oil industry.

Robert Crabbe served for 16 years in the RN as a Warfare Officer and Navigation Specialist. Since leaving the RN, Rob has been involved in ship design and HF projects for the MOD and international navies whilst working for BMT Defence Services Ltd. He is now the Customer Support Manager for SRT Ltd.

Mathew Wood is a Training Specialist at BMT RCL. His experience includes 10 years in RN uniform, initially in the Fleet Air Arm and later as a Submariner. He is responsible for supporting the development of the emerging Submarine Training Enterprise Programme and Under Water Training Programme.

Dave Collins is a HF Psychologist at the University of Central Lancashire. He has over 30 years' experience in human performance, coaching and teaching, training, and accreditation/qualification for various environments including sport, business, military/emergency services and adventure education.

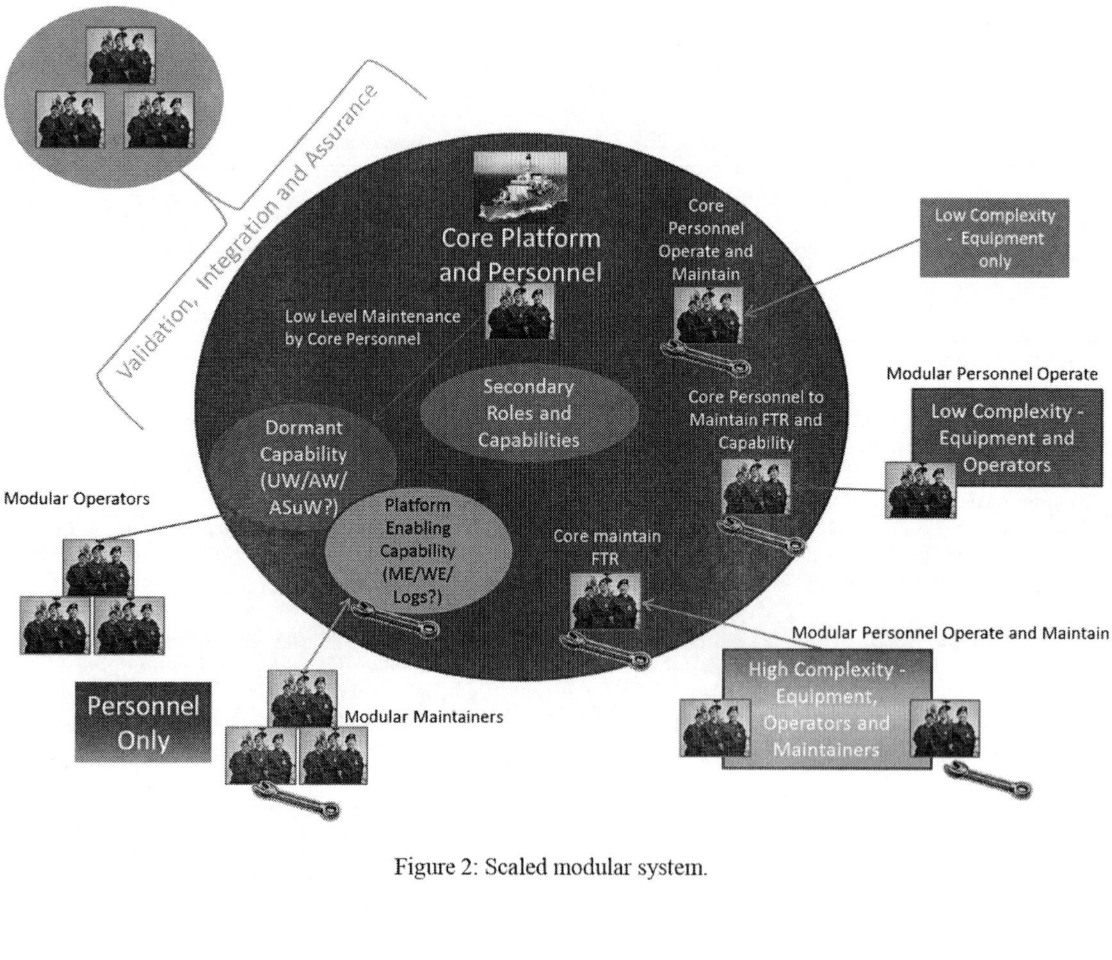

Figure 2: Scaled modular system.

RISK ASSESSMENT MODEL FOR SHIP COLLISIONS IN CONSIDERATION OF HUMAN FACTORS

T Shinoda and Department of Marine Systems Engineering, Kyushu University, Japan
K Uryu, Human Resource Group, NYK Line, Japan

SUMMARY

Most marine accidents related to vessels are the result of human factors, such as missed timely detection, judgment errors and incorrect operation of functional systems for safe navigation. In recent years, collision accidents are occurring frequently in the seas off the coast of Japan, which have been reported widely in the media. These accidents convey a negative image of marine transportation and needs to be addressed for improvements. In this paper, we examine the application of the Event Tree Analysis model and the Bayesian network model for accident occurrence modeling, which can reveal the causal relationships between human factors in maneuvering vessels and ship collisions. Then we create risk analysis models for the collision accidents based on Formal Safety Assessment (FSA) approved by IMO in 2002.

These are applied by analyzing the records of adjudication of marine accident inquiries of ship collisions between fishing vessels and cargo vessels mainly.

1. INTRODUCTION

Most marine shipping accidents related to vessels are the result of human error, such as mis-detection, mis-judgment or mis-operation in the use of functional systems that are intended to contribute to safe navigation.

In recent years, collisions involving fishing vessels and cargo vessels in the seas close to Japan have increased, in the seas close to Japan, and they have receiving greater media coverage.

The annual report released by the Japan Marine Accident Tribunal (JMAT) in 2014 shows that collisions are the most common (28%) of all marine casualties. The report also shows that the ratio involving fishing vessels is the highest (34%) of all vessel types and that the loss of life is the highest (29%), within the fishing vessel category [1].

Due to their high occurrence frequency and enormous damage, the collisions between fishing and cargo vessels are the subject of the research reported in this paper and a risk assessment is proposed in accordance with Formal Safety Assessment (FSA) procedure approved by IMO in 2002 [2].

In our previous research, first, we used the records of judgment of the marine accidents that the Marine Accidents Inquiry Agency Japan made public to extract data for human-related factors and categorize the data according to facts or the human factors involved in the collision. Second, we created a database of collisions between fishing and cargo vessels. Third, we developed a risk analysis methodology on the basis of the probability of collisions, which are taken the ship traffic data and estimated by combining the analyses of marine casualty data and the navigation track data obtained by examining the ship traffic survey in Kanmon Channel in Japan for 72 hours continuously in 2006. Event tree analysis (ETA) was applied to evaluate the collision risk. Finally, we examined the data to evaluate the risk control options (RCOs) using the developed methodology [3].

In this study, by extending our previous research, we examine the application of Bayesian network (BN) model for evaluation of the effectiveness of the RCOs as safety measures to reduce the collision risks, and show effectiveness of two risk analysis model such as ETA model and BN model.

In this paper, types of vessels defined as "cargo vessel" are general cargo ships, cement carriers, container carriers, car carriers, bulk carriers, oil tankers, LNG/LPG carriers, passenger ships, car ferries, tug boats and pusher vessels. These categorizations are based on the JMAT annual report.

2. FLOW OF RISK ASSESSMENT

The proposed risk assessment is in accordance with the FSA procedure approved by IMO in 2002. In this research, human factors are considered in connection with the FSA. The proposed risk assessment consists of five steps: identification of hazards, risk analysis, generation of RCOs, assessment of cost, and benefit and recommendation of the decision maker. The flow of the proposed risk assessment is shown in Figure 1. The details of each step are described as follows:

(Step 1) Identification of hazards:
To identify the hazards associated with human factors in the ship's maneuvering, a collision database is constructed. Moreover, to analyze the hazard associated with traffic volume, a ship traffic survey was conducted and data describing harmful situations were extracted.

To configure the collision database, the hazards attributable to human factors relating to collisions are identified using the records of the adjudication of marine accident inquiries, application of the variation tree analysis (VTA) method, and definition of maneuvering

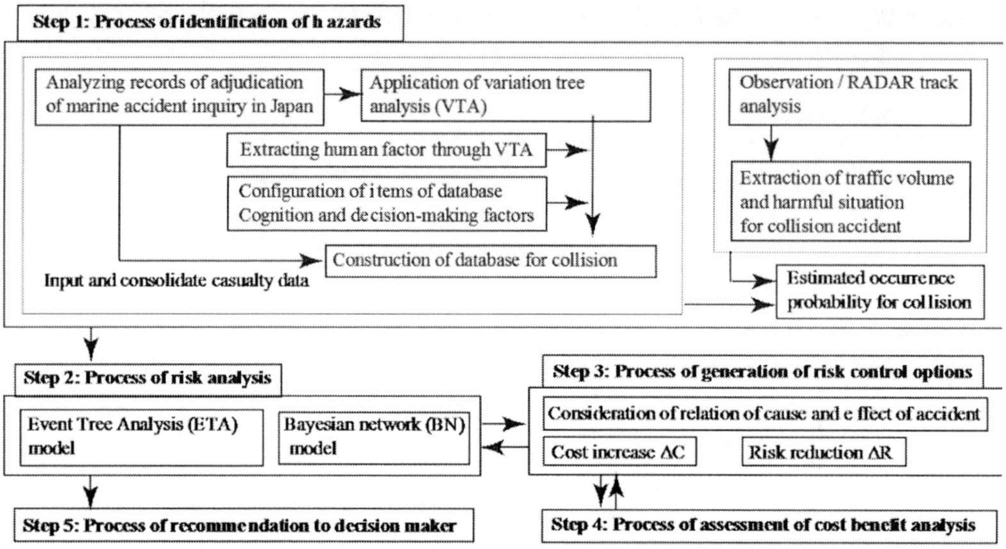

Figure 1: Flow of proposed method of safety assessment for collision accidents.

processes based on human cognitive information processing in an emergency situation.

On the other hand, for the ship traffic survey, we used the navigation tracking data obtained for all vessels by marine radars, automatic identification system (AIS), and visual observation.

The trial calculations of occurrence probabilities of collision accidents were performed using data obtained from the constructed database and the ship traffic survey. The hazard of collision is estimated by the use of occurrence probabilities of collision.

(Step 2) Risk analysis
Risk analysis model applied by ETA model and BN model is created for ships collisions to elucidate a causal relationship with human factors in maneuvering and analyze the ship collision risk.

(Step 3) Generation of risk control options
RCOs to prevent collision accidents are generated from the results of the database analysis. Safety measures such as improvement of the radar reflector, shapes with high level visibility at night time, collision warning equipment, and installation of AIS are considered practical RCO examples in this study, and the effectiveness of each RCO is analyzed by the proposed risk analysis model for collision accidents.

(Step 4) Assessment of cost benefit
The cost benefit assessment is an important step in order to decide to implement the RCOs in ship operation. Trial cost benefit analysis for RCOs is considered through gross cost of averting fatality (GCAF) in the FSA process. The risk reduction effect by risk analysis model and cost evaluation through step 3 are used in the calculation of GCAF.

(Step 5) Recommendation of decision maker
Recommendations are discussed and four steps are presented to the relevant decision makers in the relevant section of the IMO.

This study examined the feasibility of steps 1-4 while providing a trial risk assessment for ship–ship collisions between fishing and cargo vessels that are related to typical human factors.

3. SITUATION ON COLLISION BY DATABASE ANALYSIS

3.1 VARIATION TREE ANALYSIS

VTA is a method of establishing measures that can be applied to reported accidents to extract "Variation Factors" from the data. Variation factors depart from usual procedures. Speculative factors are excluded in this method. The extracted variation factors are considered as causes of accidents and arranged sequentially to show their flow. The countermeasures against accidents are investigated by reducing the flow of variation factors[3].

The sequence of variation factors is shown in tree format and variation factors are linearly arranged as a function of time. In the case of collisions, vessels' courses and their relative angles are important factors. Accordingly the visualized variation tree shown in Figure 2 is proposed to determine the progression of the factors contributing to the accident and to help understand the sequence of variation factors at each moment in crossing

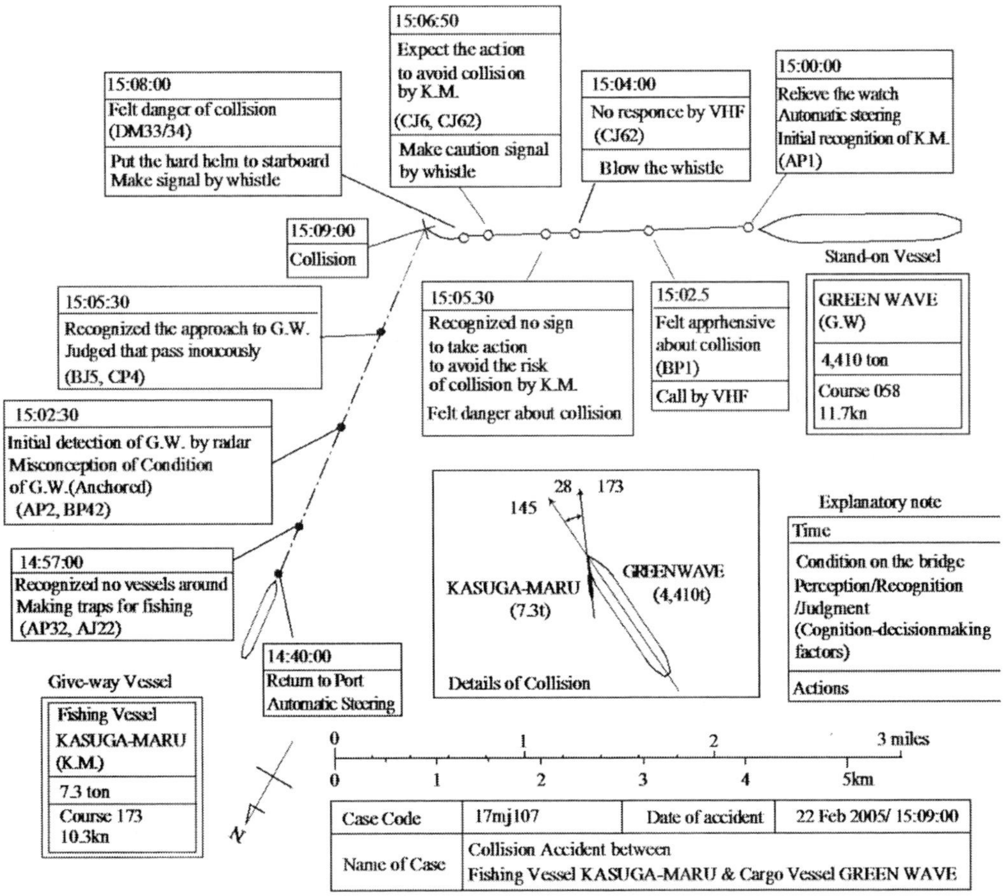

Figure 2: Visualization of collision accident by VTA.

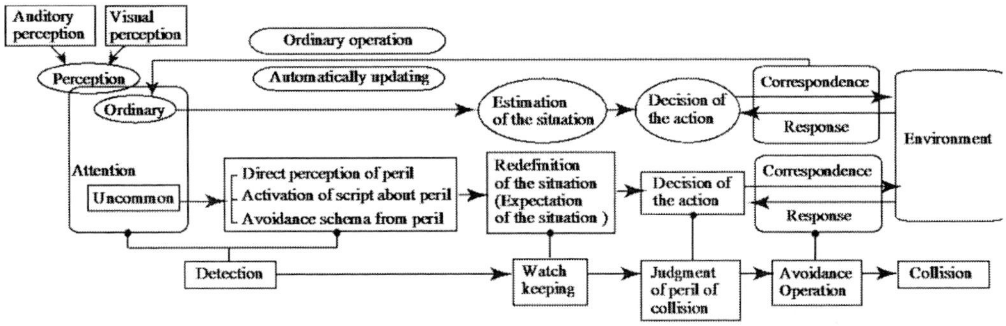

Figure 3: Model of human information processing on maneuvering.

situations. VTA has been applied to 100 records acquired from judgments rendered by the data of JMAT from 2001 to 2008[1].

3.2 MARINE CASUALTY DATABASE

The marine casualty database was constructed with human variation factors extracted from the visualized VTA and the facts obtained from the records of adjudication of the JMAT, such as accident attribution (place, time, season, weather, sea condition, visibility), vessel attribution (type, length, gross tonnage, speed, course, bridge resources, roles at crossing situation (give-way or stand-on), and other factors.

The total database consists of 217 cases that were judged in Kanmon Channel from 1995 to 2008. The items in the database are considered for the ship's maneuvering process based on human cognitive information processing in emergency situations that is defined as perception of an unusual situation, redefinition of the

Table 1: Code of cognition and decision-making factors on collision accident database.

Situation	Code	Sub-code	Factor
Situation A	AP1		Success of detection
	AP2		Success of detection by radar
Detection of opponent vessel	AP3		Failure of detection
		AP31	Distraction by other vessels
		AP32	Distraction by other works
		AP33	Improper watch
		AP34	Sleeping
		AP35	No watch at anchoring
	AJ1		Shipping route
	AJ2		Judge as safe situation
		AJ21	Judge from experiences
		AJ22	Detect no vessels by radar
		AJ23	Recognize no vessels
		AJ24	Expect other vessels action
		AJ25	Follow brother vessels
	AJ3		Non compliance of regulations
		AJ31	Improper watch assignment
		AJ32	No report on other vessels
		AJ33	Improper watch
		AJ34	No attention to blind zone
	AJ4		Fatigue
		AJ41	Feel sleepiness
		AJ42	Fatigue
		AJ43	Posture during watch (sitting...)
Situation B	BP1		Watchkeeping
Watch keeping	BP2		Watchkeeping by radar detection
	BP3		Failure of detection
	BP4		Insufficient watchkeeping
		BP41	Distraction by other vessels
		BP42	Distraction by other works
		BP43	Postponent of decision
	BJ1		Shipping route
	BJ5		Misjudge on passing innocuously
		BJ51	Misjudgment on situation
		BJ511	Judge from cursory inspection
		BJ512	Improper watch assignment
		BJ52	Misjudgment on crossing situation
		BJ53	Take the action to avoid collision
	BJ6		Expect other vessel avoidance action
		BJ61	Role by regulations
		BJ62	Take action to encourage other vessels avoidance action
		BJ63	Other vessel type is small vessel
Situation C	CP3		Failure of detection
	CP4		Insufficient watchkeeping
Judgment of peril of collision	CM1		No action to avoid collision
	CJ6		Expect other vessel avoidance action
		CJ61	Role by regulations
		CJ62	Take action to encourage other vessels avoidance action
		CJ63	Other vessel type is small vessel
		CJ64	Judge from experience (Small vessel avoidance action)
		CJ65	Expect other vessel action
		CJ66	Under fishing operation
		CJ67	Failure of judgement on other vessel condition
	CJ7		Failure of judgement on situation
		CJ71	Failure of judgement (late action)
	CJ8		Improper watch assignment
		CJ81	Expect co-watchers judge
	CM2		Failure of avoidance action
		CJ51	Failure of judgment
Situation D	DM1		No avoidance action
Avoidance Operation		DP3	Failure of detection
		DP4	Insufficient watchkeeping
		DM11	Unable to take action due to fishing operation
		DM12	Notice at close range, but unable to take action
		DM13	Expect other vessel action
		DM14	Tie down
	DM2		Improper/Insufficient avoidance action
	DM3		Action at close range
		DM31	Engine operation
		DM32	Steering operation with signal
		DM33	Steering operation without signal
		DM34	Reminder signal (whistle/light)
		DM35	Warning signal

situation, decision of the action, and correspondence to the environment, as shown in Figure 3.

Following such information processing, the flow of ship's maneuvering process is defined using four major categories: detection, watch keeping, judgment of peril of collision, and avoidance action. Finally, the influences of human factors on the collision are reconstructed for the four major categories, and the details of the categories are summarized as cognition and decision-making factors on collision accident in Table 1[4].

3.3 RESULTS OF DATABASE ANALYSIS

Figure 4 shows an example of the analysis of the constructed collision database. On the part of the fishing vessel, 69% of detection failures (AP3) are due to distraction by other tasks (AP32, 37%) and improper lookout (AP33, 53%). Even at the phase of taking measures for avoidance of the collision, 12% of insufficient watch is due to the distraction by other tasks that have been continued, leading to failure by the fishing vessels to take avoidance actions. This analysis implies that fishing vessels do not reach the phases of judgment and action due to failure of detection of cargo vessels. At the phase of crisis conditions, 67% of fishing vessels do not take any action to avoid the collision

On the other hand, the behavior of the cargo vessels is described as follows: At the earlier phases of situation A, the fraction of detection failure is 44%. The cumulated fraction of misdetection and insufficient watch at the phase of a crossing situation is 71%. At the phase of perilous conditions, 34% of cargo vessels do not take any avoidance action due to their optimistic expectation of fishing vessels' avoidance even though they recognized the peril of collision.

It is important to emphasize the significant finding of this analysis: 68% of fishing vessels do not take any avoidance action at the phase of collision crisis conditions, whereas 60% of cargo vessels take action to avoid the collision. This result indicates that the success

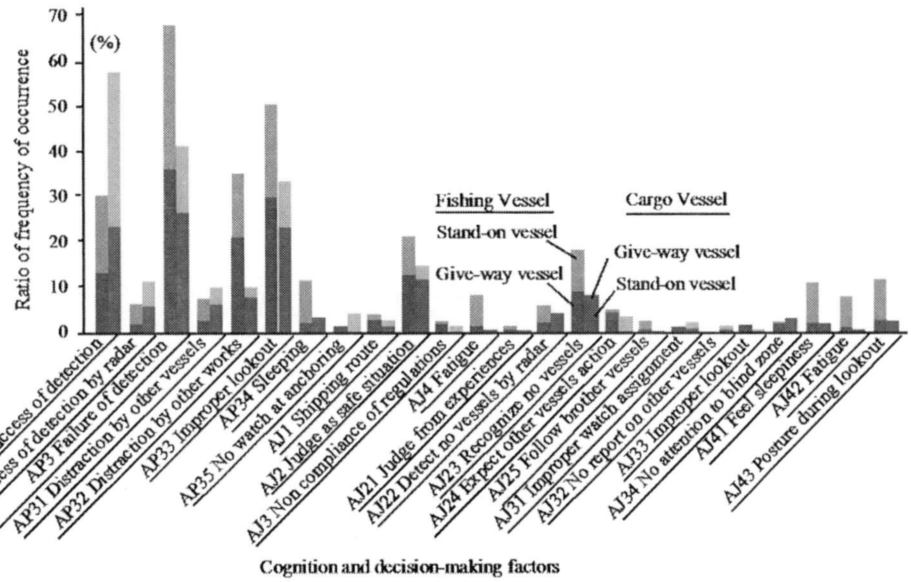

Figure 4: Frequency of cognition and decision-making factors in collision by database analysis.

Table 2: Comparison of collision probability with trafic volume.

Calculation by ship traffic volume								
Type of vessel (Unit)	Num. of vessels for 3 days (observation) (ships)	Num. of vessels (estimated) (ships / year)	Num. of accident (cases/year)	Probability of accident	Ratio of probability			
Cargo vessel	2,152	261,827	3.44	1.32×10^{-5}	1	1	-	-
Fishing vessel	404	49,153		7.01×10^{-5}	5.33	-	1	-
Calculation by num. of incidence of crossing situation								
Total	235	28,592	3.44	1.20×10^{-4}	-	9.16	1.72	-
Daytime	204	24,820	1.22	4.92×10^{-5}	-	-	-	1
Nighttime (6PM-6AM)	31	3,772	2.22	5.89×10^{-4}	-	-	-	11.96

of detection, continuous watch, and proper judgment to decide to take avoidance action are the keys for the prevention of collision accidents.

4. ESTIMATION OF PROBABILITIES OF SHIP'S COLLISIONS BASED ON SHIP'S TRAFFIC DATA

Probabilities of collision accidents are estimated by joint analysis of collision data and navigation track data. The navigation track data was obtained by analysis of the ship traffic survey conducted continuously for 72 hours in the Kanmon Channel from 0 AM on 14 November 2006 to 12 PM on 16 November 2006 [5]. The traffic data for all vessels were obtained by marine radar, AIS, and visual observation every 1–3 min, and vessel types and their sizes were observed visually or obtained using IMO ship identification data.

Here, the probability of collision accidents is defined as the ratio of collision occurrence to the amount of traffic, with the occurrence probabilities of fishing and cargo vessels calculated on the basis of traffic volume and the number of crossing situations, as shown in Table 3. The traffic data are not measured at the same sequence time, and the data are corrected by interpolation of the traffic data at every minute. Also, the crossing situation is simulated using the calculated collision points obtained by extrapolation of the traffic data at every minute.

The occurrence probability of collision of fishing vessels based on the traffic volume is 5.3 times higher than that of cargo vessels. The occurrence probabilities based on the number of crossing situation are 1.72 and 9.16 times higher than the probabilities calculated from the traffic volume for fishing and cargo vessels, respectively. The occurrence probability based on the number of crossing situation for night time is higher than that for daytime by a factor of 11.96. These results lead to the conclusion that the collision risk is much higher once a crossing situation exists and at night.

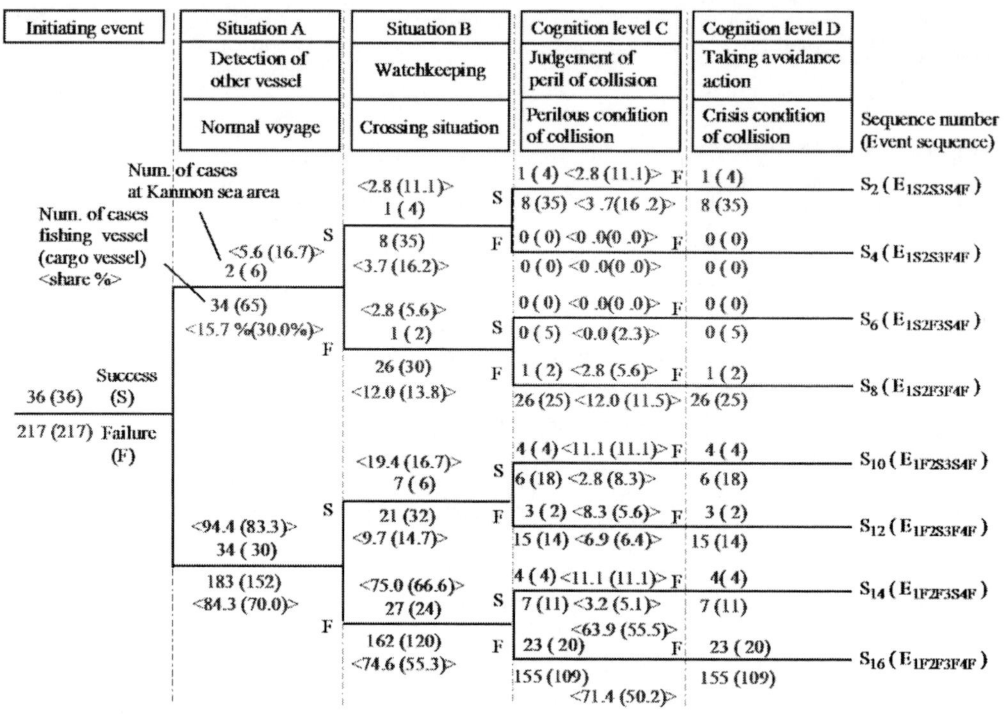

Figure 5: Event tree of collision between fishing vessel and c argo vessel.

Table 3 Definition of situation based on cognition level and prescriptive maneuvering

Denomination of situation	Level of cognition and judgement	Prescriptive maneuvering at the distance between two vessels
Situation A	Detection of other vessel at collision risk	Equal and more than 2 miles
Situation B	Watchkeeping of other vessel at collision risk	Less than 2 miles
Situation C	Judgement of peril of collision	Less than 1 mile
Situation D	Taking avoiding action	Less than 0.5 mile and close range

5. RISK ASSESSMENT BY PROBABILITIES

5.1 EVENT TREE ANALYSIS MODEL FOR SHIP'S COLLISION

Event tree analysis (ETA) is a method to visually represent a sequence of events and to estimate the many possible results that can stem from an initial event. The Event Tree bifurcates into success and failure at every phase of the sequence. "Success" is determined when the required processes for avoiding a collision situation are taken successfully and "failure" is determined when the required processes are either not taken or fail. Each bifurcation of the Event Tree relates to some safety characteristics. Probabilities are applied to success and failure at each bifurcation respectively.

The Event Tree is developed as shown in Figure 5. Four events: Detection of opponent vessel; Watchkeeping; Assessment of avoidance action and formulaic maneuvering; and Taking avoidance action are the principal events of each bifurcation. The time sequence of the event tree is expressed by four situations from Code A to Code D as shown in Table 2. The number of cases obtained from the database is indicated at each bifurcation for both the fishing vessel and the cargo vessel. Since the analysis data is taken from the records of adjudication of marine accident inquiries, the number of successes stemming from avoidance actions is unobtainable.

The developed Event Tree shows patterns of collisions, which are represented as the collision sequence number in the same figure. The most frequent sequence pattern for both fishing vessels and cargo vessels is S16, where detection of the opponent vessel at an early stage does not occur and subsequent judgment/avoidance actions are not taken. In this sequence, the failure of detection leads

Table 4: The number of classified damage according to category of collision and collisional sequence.

Totalized category		Num. of accident	Num. of death accident	Num. of death	Num. of injury accident	Num. of injury
Sequence no. S_2	F.v.	8	0	0	5	8
	C.v.	35	2 (1)	2 (3)	14	19
S_6	F.v.	0	1	1	2	6
	C.v.	5	1	1	1	1
S_8	F.v.	26	1	1	2	6
	C.v.	25	1	1	1	1
S_{10}	F.v.	6	0	0	0	0
	C.v.	18	0	0	3	3
S_{12}	F.v.	15	0	0	15	18
	C.v.	14	3	3	13	17
S_{12}	F.v.	7	0	0	15	18
	C.v.	11	3	3	13	17
S_{16}	F.v.	155	13 (1)	14 (3)	43	54
	C.v.	106	8	9	36	49
Total accident		217	14 (1)	15 (3)	67	89
At night (6P.M.-6A.M)		114	6 (1)	7 (3)	33	50
At Kanmon sea area		36	4 (1)	5 (3)	13	16
At Kanmon seaarea at night		22	3 (1)	4 (3)	9	13

Remark: The number; damage to fishing vessel ,() damage to cargo vessel
F.v.; relating to fishing vessel, C.v.; relating to cargo vessel

directly to the subsequent accident. The most frequent sequence pattern for cargo vessels is S2, where failure of the avoidance action leads to the accident. This occurs because a proper watch is not kept and subsequent recognition of the danger and effective actions are not taken once detection is established. The event tree also indicates the pattern and tendency of accidents focused on human variation factors. A similar tendency of accidents occur in all collisions and in the collisions in the Kanmon sea area based on a comparison of the incidence ratios of collision.

Table 4 shows the number of damage incidents, classified by deaths and injuries sustained, according to the category of collision and the collision sequence. For all 217 collision cases in the database, 15 persons were killed indicating that the ratio of deaths to collisions is 7%. The analysis shows that a total 89 persons were injured in 67 cases of collision, indicating a ratio of 1.32 personal injuries per collision. The most frequent pattern for both fishing vessels and cargo vessels, S16, is risk contingent to deaths and injuries. The most frequent pattern for cargo vessels, S2, is the source of the greatest incidence of deaths and injuries aboard fishing vessels.

5.2 APPLIED BAYESIAN NETWORK MODEL FOR SHIP'S COLLISION

5.2(a) Accident Occurrence Model by Applied Bayesian Network

The Bayesian network model is a convenient methodology for revealing a causal relationship and calculating the posterior probability by conditional probability. In this study, risk analysis models using Bayesian network are created for the collision accident to reveal a causal relationship with human factors in maneuvering and analyze the risk of collision using the conditional probability table calculated using the constructed collision database. The application of Bayesian network to the evaluation of RCO effectiveness is particularly effective.

The Bayesian network model for risk assessment for collision consists of three major models such as the model of ship's maneuvering, the model of collision factors, and the model of effective factors for RCOs, as shown in Figure 6. The details of the models are as follows [6].

In the model of the ship's maneuvering process, four major random variables are used. These are Situation A: sensibility, Situation B: watch keeping, Situation C: judgment of peril of collision, and Situation D: avoidance maneuvering. While in principle, many characteristics of the ship's state such as the distance, angle, vessel velocity, and ship's operation can be used to set the random variables, in this study we use the distance between the two ships involved in possible collision to set the random variables set because inquiry records include detailed descriptions of the distance between the two ships until collision.

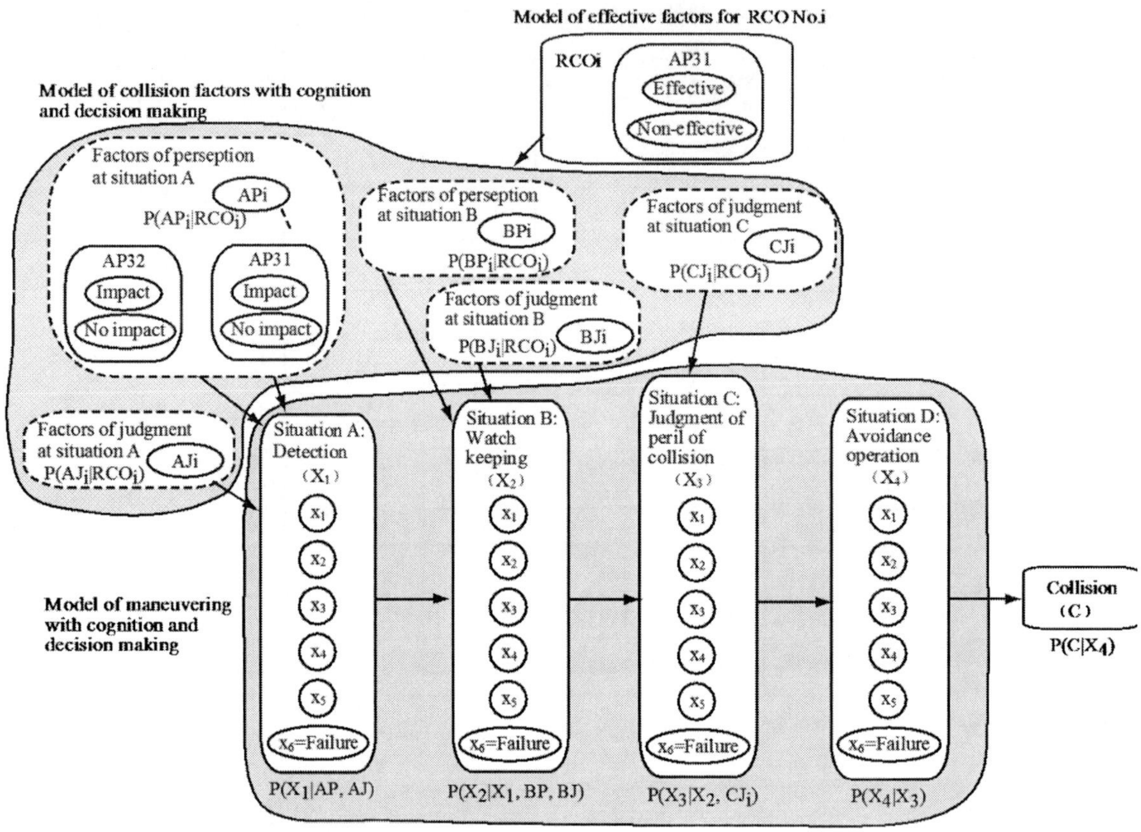

(a) Bayesian network of evaluation of RCO' s effective for collision

(b) Random variables and its state on Bayesian network

Random variables		State of random variable	
X_1	Detection	$x_1, x_2, x_3, x_4, x_5, x_6$=f	Position at detection x_j and its failure f
X_2	Watch keeping	$x_1, x_2, x_3, x_4, x_5, x_6$=f	Position at watch keeping x_j and its failure f
X_3	Judgment of taking avoidance	$x_1, x_2, x_3, x_4, x_5, x_6$=f	Position at Judgment of taking avoidance x_j and its failure f
X_4	Taking avoidance action	$x_1, x_2, x_3, x_4, x_5, x_6$=f	Position at taking avoidance action x_j and its failure f
C	Collision	x_1=f	Failure by collision
APi	Factors of perception at situation A (4 items)	fi, fn	Influance factor fi and non-influance factor fn at detection
AJi	Factors of judgment at situation A (8 items)	Do.	
BPi	Factors of perception at situation B (3 items)	Do.	
BJi	Factors of judgment at situation B (2 items)	Do.	
CJi	Factors of judgment at situation C (3 items)	Do.	
RCOi	Effective factors for RCO No.i	ef, ne	Effective ef and non-effective en

Remark: x_1: Equal and more than 2 mailes, x_2: Less than 2 mailes, x_3: Less than 1maile, x_4: Less than 0.5 maile, x_5: Close range

Figure 6: Bayesian network model for risk assessment for collision.

Table 5 shows the categories of vessel–vessel distances such as more than 2 miles, 1 to 2 miles, 0.5 to 1 mile, 0.5 mile or less, and close range. Each distance category is related to the ordinary level of cognition and judgment in the prescribed maneuvering. This model considers the distance category as the status of random variables ranging from x_1 to x_6, as shown in Table 3. All random variables and their states are listed in Figure 6.

Table 5: Definition of random valuable based on cognition level and prescriptive maneuvering.

Random variable	Distance division between two vessels	Ordinary level of cognition and judgment in the prescriptive maneuvering
x_1	Equal and more than 2 miles	Detection of other vessel at collision risk
x_2	Less than 2 miles	Watch keeping of other vessel at collision risk
x_3	Less than 1 mile	Judgment of peril of collision
x_4	Less than 0.5 mile	Taking avoiding action
x_5	Close range	-
x_6	Failure	-

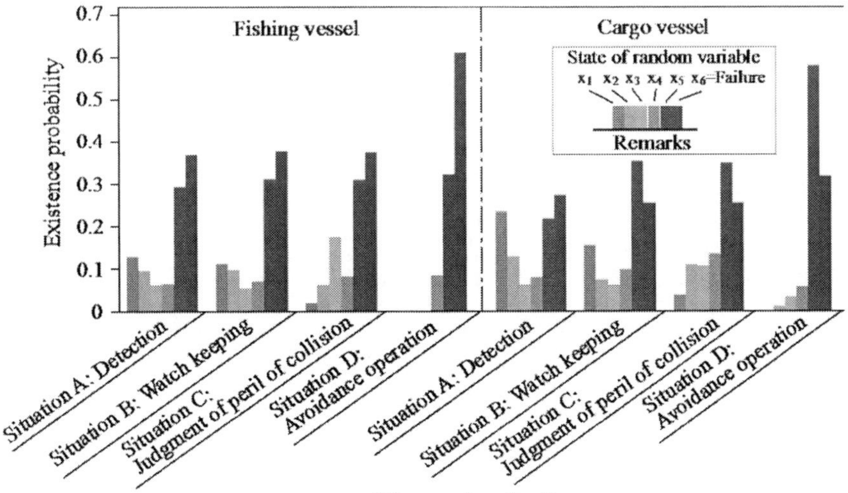

Figure 7: Calculation results of posterior probability of each maneuvering situation in collision.

In the model of cognition and decision-making collision factors, although all factors shown in Table 1 should be considered as random variables, due to the limitation of Bayesian network software, in this study only the representative factors are selected and their existence probability is calculated.

In the model of effective factors for RCO, collision factors associated with the effectiveness of RCO are reduced when the RCO functions effectively, and a new model of collision factors is then reconstructed. Calculation with the new model results in a change of existence probability; this change represents the effectiveness of the RCO. The details of the method for evaluation of the RCO effectiveness are described in APPENDIX.

5.2(b) Risk analysis of collision accidents

Figure 7 shows the existence probability values $P_{0,ik}$ at random value x_k in each state X_i. The $P_{0,11}$ that represents the existence probability value of the fishing vessel at a random value x_1 in state X_1 is 0.13. While the value of $P_{0,11}$ is high when detection (X_1) is normally performed in the state of the ship–ship distance of more than 2 mile ($X_1 = x_1$), the probability value on the detection is low for the occurrence of accidents. Even when the detection is conducted at more than 1 mile ($X_1 = x_2$), the existence probability value is calculated as a sum of $P_{0,11}$ and $P_{0,12}$ and is equal to 0.22; however, this is low. The existence probability values on detection suggest that the appropriate on-board detection was not operating when accidents occurred. On the other hand, $P_{0,15}$ is 0.29 when the detection is abnormally carried out in the state of close range ($X_1 = x_5$) and $P_{0,1f}$ is 0.37 when detection failed. The sum of these existence probabilities is 0.66, which is extremely high. This also suggests that the delay of detection is correlated with the occurrence of accidents.

The existence probability values in each situation of watch keeping, judgment of peril of collision, and avoidance operation are similar to the value of the situation of detection, indicating that the operational delays appear in each situation.

Finally, the sum of the probability values of $P_{0,44}$, $P_{0,45}$, and $P_{0,4f}$ increases to 0.91, suggesting that delay or failure of avoidance operation appear for occurrence of accidents.

On the other hand, in the case of cargo vessels, the existence probability value on the situation of detection is 0.37 when the detection is carried out by $X_1 = x_2$. This

Table 6: Estimated cost and impacts to cognition and decision-making factors on each RCOs.

RCO No.	Measures	Effectvity of measures	Expected impact to cognition and decision-making factors	Expected cost (USD)
RCO #1	Installation of improved radar reflector	Perception and judgement for cargo vessel	AP31, AP33, AJ22, AJ34 BP41, BP43, BJ511	350
RCO #2	Installation of bright instrument	Perception and judgement for cargo vessel at night	AP31, AP33, AP34, AJ23 BP41, BP43	450
RCO #3	Installation of approach warning system	Perception and judgement for fishing vessel	AP31, AP32, AP33, AP34, AJ33, AJ34, BP41, BP42, BP43, BJ511	350
RCO #4	Installation of scale-downed AIS	Perception and judgement for cargo and fishing vessel	AP31, AP32, AP33, AJ22, AJ33, AJ34, BP41, BP42, BP43, BJ511	2,500

value is better than that of the fishing vessels. However, the appropriate watch keeping and judgment after detection operations are not conducted within a range sufficient for avoidance of collisions. Finally, the existence probability value of avoidance operation increases to 0.58 at close range.

This trend suggests that earlier detection on cargo vessels is not effective for watch keeping and judgment of peril of collision.

6. RCO EFFECTIVE EVALUATION

6.1 RISK CONTROL OPTIONS (RCOS)

According to the trend exhibited by the probability values as described in the previous section, RCOs for supporting detection are effective for collision prevention. The effectiveness of RCO is evaluated by the proposed method using the RCO effectiveness index. In this study, due to easy installation, the following equipment is considered for RCOs for fishing vessels and its effectiveness is evaluated.

RCO #1) Improvement of radar reflector:
The collision occurs even when fishing vessels install a radar reflector. This result suggests that it is necessary to improve exiting radar reflectors to be more highly reflective so that cargo vessels can easily detect the fishing vessels. It is expected that installation of a highly reflective radar reflector on the fishing vessels will make the fishing vessels effectively visible for cargo vessels.

RCO #2) Setting highly visible fishing shape at night:
Collision accidents at night are high probability events. Setting highly visible fishing shapes on the fishing vessels is effective at night.

RCO #3) Setting approach warning apparatus:
Detection of the fishing vessels can be effectively supported using an approach warning apparatus on the fishing vessels to make it known that the cargo vessels are approaching.

RCO #4) Setting class-B AIS:
According to regulation 19 of IMO SOLAS Chapter V, the regulation requires AIS to be installed aboard all international ships of 300 gross tonnage and upwards engaged on international voyages, all cargo ships of 500 gross tonnage and upwards not engaged on international voyages, and all passenger ships irrespective of size [7]. The installation of the class-B AIS on all cargo vessels and fishing vessels is considered as RCO. The system that provides the current information such as ship name, speed, and course to monitor other vessels are the minimal requirements for this system. The system also must be inexpensive and easy to install on the fishing vessels.

Table 6 shows the list of estimated costs and impacts on the cognition and decision-making factors on RCOs. For example, for RCO #4, AP31, AJ22, AJ23, AJ33, and AJ34 that are described in Table 3 are cognition and decision-making factors defined in Table 1, and it is assumed that these failure factors are reduced when AIS is introduced in the fishing vessels.

In the ETA model, the effectiveness of each RCO are estimated to each cognition-decision making factors in Table 1 with the risk reduction rate as the impact of the total number of accidents. The introduced reduction rate 50% on each recognition level during navigation is adopted for the estimation of the efficiency of each RCO.

Also in the RCO effectiveness model in BN model, these impact items are considered as the elements of SI RCOi described in the Method of Evaluation of Effectiveness of RCOs section in APPENDIX. In calculation of efficiency of each RCO, these factors are eliminated from the original Bayesian network model for risk assessment. The new network is reconstructed and existence probabilities are calculated.

Table 7: Calculation results of GCAF on RCO.

RCO No.	Expected cost, ΔC	Expected cost per year, ΔC	ETA model Total risk reduction rate	BN model Total RCO effectiveness index, ΔR	GCAF (total cost basis)	GCAF (cost per year basis)	Ratio of GCAF to WTP
Unit	(USD)	(USD/year)	(%)	(%)	(USD)	(USD/year)	
RCO #1	350	36	24.1	24.1	1,454	151	0.802
RCO #2	450	47	22.0	28.9	1,555	162	0.858
RCO #3	350	36	34.3	27.7	1,263	131	0.697
RCO #4	2,500	260	58.4	61.0	4,095	425	2.259
WTP-basis	1,105	115	-	61.0	1,810	188	1.000

6.2 EVALUATION OF COST BENEFIT OF RCOS

GCAF, an index of cost effectiveness in FSA, is used for evaluation of cost benefit. The formula of GCAF is defined as follows [2]:

$$GCAF = C / R \qquad (1)$$

Where ΔC is the cost increase factor and ΔR represents the risk reduction by introducing RCOs. In this study, ΔC is set to the price of RCO including installation fee and ΔR is set to the total RCO effectiveness index. The cost effectiveness is shown by the ratio of both factors, and RCO is more cost effective when for smaller GCAF values. According to the definition of risk reduction by a RCO is estimated as the reduced number of loss of life in the process of evaluation by GCAF. However, the collision accidents between fishing vessels and cargo vessels are directly linked to loss of life for fishing vessels. In this research, all accidents including property damage accident, fatal and injury accident should be analyzed.

When the cost of RCO is considered, it is necessary to investigate the acceptance of payment by fishermen as ΔC. In order to obtain the opinion of fishermen on safety costs, we carried out a questionnaire survey about the present operation, an unsafe situation, the amount that the fishermen are willing to pay, and the duration of time to pay for safety improvements. The interview were conducted over 13 times at the Kanesaki fishing harbor in Fukuoka prefecture from October to December, 2009, and 53 survey responses were obtained. According to the survey responses, the amount of payment that the fishermen were willing to pay (WTP) was around 115 USD per year, and the average duration for payment was 9.65 years. Therefore, the total amount of payment provided by the fishermen is estimated to be around 1,105 USD [4].

Table 7 shows the calculated results of GCAF on each RCO. In this table, a comparison of ΔR by calculation of ETA model and BN model are included. These values ΔR resemble closely in the calculation with reduction rate 50% on ET model, and particularly the value of ΔR of RCO #4 has the highest effectiveness.

In the calculation of GCAF on WTP-basis, the total RCO effectiveness index on RCO #4 are used, which is the highest effectiveness index in RCOs, and WTP by fishermen as a reference for comparison.

According to the cost comparison data, since cost on RCO #1, #2 and #3 are less than cost on WTP-basis, fishermen can accept the introduction of these as safety measures. However, since the total RCO effectiveness indices on RCO #1, #2, and #3 are much smaller than the index on RCO #4, fishermen hesitate to accept that the introduction of these RCOs is necessary to improve safety performance.

When each GCAF on RCO is compared with GCAF of WTP-basis, because the GCAF on RCO #1, #2, and #3 is less than the GCAF on WTP-basis, fishermen can accept the introduction of these as safety measures. On the other hand, since the GCAF on RCO #4 has a large difference with that of WTP-basis and the GCAF difference is greater than a factor of two, fishermen hesitate to introduce RCO #4. If the price of RCO #4 is reduced by at least a half, there is a high possibility that RCO #4 will be accepted.

Thus, the possibility of acceptation of RCO by the fishermen can be evaluated from comparison of GCAF based on WTP-basis.

7. CONCLUSION

A risk assessment is proposed in accordance with the FSA procedure and is applied to the problem of collisions between fishing and cargo vessels. Through the application of the model to practical vessel operations, a risk analysis model for ship collisions using Event Tree Analysis model and Bayesian network model, and a method of evaluation of effectiveness of RCO are proposed. These analytical methods are then applied to the problem of collisions between fishing and cargo vessels. Cost benefit assessment based on GCAF in FSA is proposed and the possibility of acceptance of RCO by the fishermen can be evaluated from comparison of GCAF based on WTP-basis.

Note that the proposed methodologies are efficient for providing risk assessment for the problem of collisions.

8. REFERENCES

1. Japan Marine Accident Tribunal, *The report of Marine Accident Inquiry*, for example, http://www.mlit.go.jp/jmat/saiketsu/saiketsu.htm, (accessed on 10th Aug. 2016)

2. International Maritime Organization (IMO). IMO MSC-MEPC. 2/ Circ. 12, 2013

3. T Shinoda, Y Tamura, Risk Assessment of Collision Accidents at Sea in Consideration of Human Factor -Application of Collision between Fishing Vessels and Cargo Vessels-, *Journal of Institute Navigation, vol. 124, pp.11-19*, 2011.

4. Y Tamura, T Shinoda, Safety Assessment with Marine Casualty Data on Collision Accidents between Fishing Vessels and Cargo Vessels, *Proceeding of ISOPE (The International Society of Offshore and Polar Engineers), pp.641-648*, 2009.

5. Kyushu Regional Development Bureau of Ports and Harbor Bureau of Ministry of Land, Infrastructure, Transport and Tourism The ship traffic report in Kanmon Channel, 2007.

6. T Shinoda, K Shimogawa, Y Tamura, *Applied Bayesian Network Risk Assessment for Collision Accidents between Fishing Vessels and Cargo Vessels, Journal of Institute Navigation, vol. 127,pp.165-174*, 2012.

7. International Maritime Organization (IMO), IMO MSC-MEPC. 2/ Circ. 12, 2013. http://www.imo.org/OurWork/Safety/Navigation/Pages/AIS.aspx, (accessed on 10th Aug. 2016)

8. Y Tamura, T Shinoda, Hearing Investigation to Fishermen on Collisions with Cargo Vessels, *Journal of Institute Navigation, vol. 128, pp.81-88*, 2013.

9. Y Motomura, Technology of Bayesian network, *Tokyo Denki University Press*, 2006.

9. AUTHORS BIOGRAPHY

Takeshi Shinoda is a professor of Kyushu University, Japan. He received D.Eng., M.A. and B.A. in naval architecture from Kyushu University. His past academic posts include a guest professor at Sao Paulo University, Brazil and a guest lecturer at Hiroshima University, Japan. His main research field is equipment engineering in naval architecture, safety engineering, sea transportation including containerized transportation and its facility, and environmental engineering concerning ship. He has received the award for excellent paper from the Japan Society of Naval Architects and Ocean Engineering in 1991 and 2010, Lifetime Achievement Award for the Japan Society of Naval Architects and Ocean Engineering in 2014, and the award for business-academia collaboration research from Japan Cabinet Office in 2014.

APPENDIX

Method of Evaluation of Effectiveness of RCOs by Applied Bayesian Network

(1) Setting of CPT and impact of RCO

In the beginning of calculation of Bayesian network, the conditional probability tables (CPT) for the model of ship's maneuvering process are created using the constructed collision database. These are given as follows:

$$\begin{aligned} &P(X_1 | AP_{i1}, AJ_{i1}) \\ &P(X_2 | X_1, BP_{i1}, BJ_{i1}) \\ &P(X_3 | X_2, CJ_{i1}) \\ &P(X_4 | X_3) \\ &P(C | X_4) \end{aligned} \quad \text{(A1)}$$

The tables relate the variables APi, AJi, BPi, BJi, and CJi that are defined in the list of random variables in Figure 6 and described in Table 1.

For the evaluation of the effect of safety improvement of RCO we define a set of safety impact factors on RCO as SI_{RCOi}. The elements of SI_{RCOi} consist of effective factors for all collision factors. For example, when the RCO influences the collision factor on APi, the CPT of APi takes the value 1. Otherwise, it takes the value 0. The formula for the CPT is expressed as follows:

$$P(AP_{i1} | RCO_i) = \begin{matrix} 1 \ \ when\ AP_{i1} \quad SI_{RCO_i} \\ 0 \ \ when\ AP_{i1} \quad SI_{RCO_i} \end{matrix} \quad \text{(A2)}$$

(2) Estimation of State of Maneuvering

Let $P_{RCOi,jk}$ denote the estimated existence probability after taking RCO No.i based on each situation X_j on ship's maneuvering and each distance between two vessels xk. The formulae for $P_{RCOi,jk}$ are given as follows:

$$
\begin{aligned}
P_{RCO_i,1k} &= P(X_1 = x_k \mid C = f,\ AP_{i1} = fi,\ AJ_{i1} = fi) \\
P_{RCO_i,2k} &= P(X_2 = x_k \mid C = f,\ BP_{i1} = fi,\ BJ_{i1} = fi) \\
P_{RCO_i,3k} &= P(X_3 = x_k \mid C = f,\ CJ_{i1} = fi) \\
P_{RCO_i,4k} &= P(X_4 = x_k \mid C = f)
\end{aligned} \tag{A3}
$$

The values of existence probability of each situation on ship's maneuvering are calculated using these conditional probabilities and the safety state of maneuvering can then be obtained. In this study, the6"Byonet" software is used for Bayesian network calculation [9].

(3) RCO Effectiveness Index

The effectiveness of an RCO can be obtained by calculating the difference in existence probability before and after RCO measures.

Let $P_{0,jk}$ denote the existence probability before taking the RCO; the matrix for evaluation on dominance D of $RCO_{i,jk}$ that shows the existence probability differences is defined as follows:

$$
D_{RCO_i,jk} = \begin{cases} M_{D,jk}(P_{RCO_i,jk} - P_{0,jk}) & when\_\mathrm{sgn}(P_{RCO_i,jk} - P_{0,jk}) = M_{D,jk} \\ 0 & when\_\mathrm{sgn}(P_{RCO_i,jk} - P_{0,jk}) \quad M_{D,jk} \end{cases} \tag{A4}
$$

where "sgn" is the signum function, with values taking 1 or 0 or −1 for positive and 0 and negative values in the brackets of sgn, respectively. Let $M_{D,jk}$ denote the matrix defining the dominance of $P_{RCOi,jk}$ and given as follows:

$$
M_{D,jk} = \begin{matrix} 1 & 1 & -1 & -1 & -1 & -1 \\ 1 & 1 & 1 & -1 & -1 & -1 \\ 1 & 1 & 1 & 1 & -1 & -1 \\ 1 & 1 & 1 & 1 & -1 & -1 \end{matrix} \tag{A5}
$$

where the line of matrix represents each situation X_j of the ship's maneuvering process and the column represents each category of the distance between two vessels x_k. Each element of $M_{D,jk}$ is assigned either 1 or -1 to represent the effectiveness of the ship's maneuvering. Here, the elements of $M_{D,jk}$ are set through an interview survey of the captains of the cargo ships.

For example, the first line of matrix $M_{D,jk}$ treats the situation of X_1 showing detection, and detection on ship' s maneuvering generally should be finished earlier. The elements of $M_{D,jk}$ expression take the positive value 1 until x_2 to evaluate the effectiveness of detection by the safety improvement on RCO.

Furthermore, other elements after x_3 take the negative value -1 because the difference of $P_{RCOi,jk}$ and $P_{0,jk}$ is negative, changing the negative sign to evaluate $D_{RCOi,jk}$.

Let $D_{RCOi,j}$ denote the dominance index for the situation that expresses the effectiveness of RCO in each maneuvering process. $D_{RCOi,j}$ can then be calculated using Eq. (4) as the sum of $D_{RCOi,jk}$ over x_k, given as follows:

$$
D_{RCO_i,j} = \sum_{k=1}^{n_k} (D_{RCO_i,jk}) \tag{A6}
$$

Let D_{RCOi} denote the RCO dominance index that expresses the total effectiveness of RCO. Using Eq. (6), D_{RCOi} is then calculated as the sum of $D_{RCOi,j}$ over j, given as follows:

$$
D_{RCO_i} = \sum_{j=1}^{n_s} (D_{RCO_i,j}) / n_s \tag{A7}
$$

The dominance index must be able to evaluate the total effectiveness of RCO. Let D_{all} denote total effectiveness when the RCO has effectiveness on all factors. The value of RCO effectiveness V_{ROCi} can then be defined as follows:

$$
V_{RCO_i} = D_{RCO_i} / D_{all} \tag{A8}
$$

Since we focus on the collision of fishing and cargo vessels, in this study, the total RCO effectiveness index for both vessels denoted as TV_{RCOi} can be calculated using following formula.

$$
TV_{RCO_i} = (VF_{RCO_i} + VC_{RCO_i}) / 2 \tag{A9}
$$

where VF_{RCOi} and VC_{RCOi} denote V_{RCOi} of the fishing and cargo vessels, respectively.

HUMAN FACTORS IN SHIP DESIGN; IMPLICATIONS FOR SHIP REPAIR & SHIPBUILDING WORKERS

P.M. Hettiarachchie, Colombo Dockyard Plc, Sri Lanka

SUMMARY

Ship Repair and Ship Building is an industry where a significant number of serious accidents, fatalities as well as occupational diseases are reported in around the world. So, it is important to understand the root causes for these accidents, occupational illnesses and diseases because Human Capital is the most valuable asset in any industry. According to researchers, 80% of industrial accidents are due to the Human Factor (HF). So, the aim of this paper is to discuss to what extent this HF has been taken into consideration in design stage of ships with respect to its construction and repair and how it impacts the employees in Ship Repair and Ship Building. For this study, the researcher has selected some case studies pertaining to shipyard accidents and near misses reported in Sri Lanka during the 15 years period from year 2000 to 2015 along with previous findings of researchers to illustrate how the design implications affected those shipyard accidents and possible countermeasures to prevent recurrence.

1. INTRODUCTION

The critical nature of Ship Repair and Ship Building had been formally addressed decades ago in 1963 at its 154th Session of the Governing Body of the International Labour Office and adopted the proposal for drawing a Code of practice on Occupational Health & Safety for Ship Repair & Ship Building industry[1]. Ever since, more than five decades have passed and many technical evolutions have taken place but still the fatality and accident rates in this industry is significantly high. According to Beck (1992)[2], "the knowledge is spreading that the sources of wealth are 'polluted' by growing 'hazardous side effects'. This is not at all new, but it has remained unnoticed for a long time in the effort to overcome poverty". However, with the development of society this paradigm has shifted and the current society is not ready to accept the risk as it was in the past and hence proper attention has to be paid to prevent accidents in shipyards. Diversification of the media, especially through social media, can tarnish the image of any reputed company overnight following an accident taken place in it; so in addition to a safety concern accidents can also prove to be a dangerous business threat.

The other fact is, researchers have proven that the Human factor contribution for accidents is around 80% [3]. This percentage is reasonably in line with the outcome of the accident analysis of leading local shipyards [4]. According to Professor Kuo[5], humans contribute around 60% directly and 30% indirectly to failures. Meanwhile, as cited by professor Robert Bea, 80% of the catastrophic failures are rooted in Extrinsic Uncertainties (Human and Organizational performance, knowledge acquisition and utilization) and only 20% of the failures are rooted in Intrinsic Uncertainties. He further shows that approximately 80% of these failures develop during the phases of operating and maintenance of the system. Most importantly, according to Bea, more than 60% of these failures develop during the design phase [6]. Therefore, due consideration should be given to HF during the design stage in order to prevent possible accidents during construction, operation, maintenance and dismantling of ships.

2. COMMERCIAL PRESSURE IN SHIP REPAIR & SHIPBUILDING

Ships are an immense financial commitment to the parties involved, where they seek to generate a tangible output at the earliest possible instance. Therefore, it is natural for ship owners to pressurize shipyards to complete the shipbuilding projects in a shortest time span possible. In the case of ship repairs, the ship owners or ship charterers do not want to have their ships laying at berths or in docks so they always persuade shipyards to complete the task at the earliest. On the other hand, ship yards are also trying their best to maximize their profits by completing a higher number of projects in a shortest time span. This urgency makes the situation worse and tends vulnerable to higher risks. Therefore, it is hard to find any time to take appropriate precautions for any omissions made during the ship design phase as far as safety of employees in shipyards are concerned, which could lead to injury or ill health to them. The current economic downturn in shipping industry together with the unusual market behaviour of Crude Oil prices also signifies this situation because shipyards are in a desperate situation in securing jobs and this immense competition does not provide any liberty for shipyards to demand for higher safety standards unless otherwise the ship owners themselves are maintaining higher safety standards on their fleets as a policy. Therefore any lapse incurred during the design stage could lead to an injury or ill health during the lifecycle of the ship. According to Reason[7] design failures is one of 11 General Failure Types (GFT) which could lead to accidents. Therefore, this paper examines the HFs which require more attention during the ship design stage and their implication to the shipyard employees giving special attention to the selected shipyard incidents, occupational illnesses and diseases discussed in previous research.

3. COMPETENCY OF SHIPYARD WORKERS

3.1 CASE STUDY

While one subcontract employee was attending to a repair on a lifeboat of a dry-docked ship, another person entered into the lifeboat and pulled the disengaging lever from inside like shifting a gear of a vehicle. The result was that the lifeboat got released from the hook and fell in to the dry-dock killing one person and causing the ignorant operator a partial permanent disability. The incident took place on 11th August of year 2000.

3.2 CAUSE OF THE ACCIDENT

The victim who pulled the lever was not aware of the function of it and obviously the possible effect and consequences of pulling it.

3.3 DISCUSSION:

Apart from the above incident, few more incidents were reported due to the lack of competencies of subcontract employees. There were three incidents during the 15 years of consideration for this paper, where the anchors of dry-docked vessels were released and fell to the dock bottom, fortunately not harming the people at the bottom. The main cause for these incidents were that the subcontract employees deployed to work with forward windlass did not have the knowledge on how the anchor is operated and its consequences, were similar to the case study discussed in 3.1.

During normal operations of ships, such machinery is handled by competent personnel who are properly trained for the particular task and frequently refreshed on necessary technical changes. Numerous simulation techniques and researches have been done to avoid onboard accidents and those were taken well into consideration in developing new standards and resolutions. However, the question of concern is whether this aspect of employee safety has been given due consideration at the ship design stage of shipbuilding and ship repair activities.

In today's world a significant number of expatriates or seasonal workers are working in shipyards. It has been assessed that 50-70% of the value added at shipyards is derived from external subcontractors and suppliers, whereas for more complex ships this figure can be as high as70-80% [8].

It is estimated that nearly 5000 Polish employees are working in the Norwegian maritime industry and 30% of these people are working in Norwegian shipyards as shipbuilders, welders and pipe fitters. Many Polish shipbuilders are also employed in Germany, the United Kingdom, Ireland, France and also Finland. The same source revealed that in Italian shipyards subcontracting is nearly 70% [9]. Likewise, Turkey shipyards are hiring 80% employees from third party companies [10]. In Sri Lanka, the leading shipyard is hiring more than 50% of subcontract employees[11].

In Sri Lanka's context, some of these subcontract employees are not properly trained and competent for their tasks. Majority of them are from rural areas and most important fact is, these workers are entrusted in handling the most safety critical work. So these subcontract employees are not in a position to understand the critical operations of high-tech machinery. Therefore in ship designing, this HF consideration with respect to semi or unskilled employees in shipyards needs to be considered.

4. UNPROTECTED OPENINGS

4.1 CASE STUDY

A ship captain of a tanker vessel fell through a deck opening on second deck level of his ship on 9th February 2000, while he was inspecting the Forepeak tank and succumbed to death due to injuries.

4.2 CAUSE OF THE ACCIDENT

Unprotected opening inside the Forepeak tank.

4.3 DISCUSSION

Unprotected deck and manhole openings are key contributing factors for shipyard accidents. In year 2001, a 26 year old ship yard engineer fell through a manhole in a similar manner to the above incident while on an inspection of scaffolding. He was fortunate to escape death but was severely injured.

In 2007 another engineer of 51 years of age got himself hurt due to an unprotected deck manhole opening in a newly built ship causing partial permanent disability.

In another incident which occurred on 8th July 2015, a subcontract employee of 22 years who was on his first day of work in the shipyard, fell into a deep Heavy Fuel Oil tank through its manhole in the main deck close to the pump room of a Tanker vessel.

During the 15 years period under consideration for this study, an annual average of 03 employees hurt themselves due to unprotected openings on-board. Most of those incidents were reportable accidents where more than 03 working days were lost while others were more critical or at times even fatal.

As discussed above, both ship charterers as well as the shipyard are highly concerned with completing the repair or construction work of the vessel at the earliest and proceed to the next charter. In such circumstances, spending additional time for erecting temporary barriers for openings or manholes is not encouraging by both parties. On the other hand, deciding the bearer of the financial aspect of this additional job could result in a stalemate. In some instances the work to be carried out inside a tank is not as much when compared to the involvement of erecting a safety fence around the opening. Therefore employees are persuaded to work around unprotected openings. But the omission of such minor aspects, yet critical, create fatal hazards throughout the lifecycle of the ship. Therefore HF consideration in providing adequate guarding around manholes and openings is crucial for all parties including ship crews, shipyard employees and the inspection officers such as third party Classification Societies surveyors.

5. TRAUMATIC SPRAINS AND CONTUSIONS DUE TO NON ERGONOMIC DESIGN

An average of more than 20% of accident leave of absence consumed by shipyard employees is due to traumatic sprains. Such sprains are mainly caused by non ergonomic designs on ships. It is understandable that ships are considerably complicated designs and it is not an easy task to address all the requirements of every stakeholder. However, with growing concern of HF and employees' safety and well-being, ship yards are facing a serious challenge in addressing those ill-treated HF at construction and repair stages.

5.1 LADDERS

The vertical ladders are quite common in most types of vessels. This is one of the two main causes for sprains where the other one is manholes. An employee detailed to work in a cargo hold has to climb up and down several times a day and this has to be continued till he completes the job he has been assigned to. Normally shipyard employees have to carry several tools and equipment, which are needed to perform their work. For example, a welder needs to carry a welding cable, chipping hammer and grinder, in addition to his personal protective equipment such as welding shield, respirator, gloves and leg guards. Therefore, using a vertical ladder carrying such equipment could be a challenging task. In such a situation, the basics of ladder safety such as three point contact or ladder angle of 1:4 (or 75^0) may not be realistic. Hence, as far as construction and repair work is concerned more attention is required in ensuring safety of ladders and stairs considering their work involvement and frequency of use.

5.2 MANHOLES AND ACCESS

Some traumatic sprains were recorded due to the arrangements of manhole access to confined spaces. Height, size and handhold are main concerns of manhole access, which cause sprains to shipyard employees. As it was highlighted in the previous section, shipyard employees have to go through these manholes several times during its construction and repair period, which makes the possibility of being injured very high. There are many confined spaces where only one manhole is available. Thus, once an employee enters in the confined space he/she cannot turn around to exit, instead he/she has to come out backwards, which is a serious challenge in small confined areas such as engine sumps. In addition to the access, service lines such as ventilation ducts, power cables and gas hoses should also go through the same manhole, which is a serious risk to shipbuilding and ship repairing employees. Disturbances such as pipelines across the access of manholes are some other reasons causing sprains to employees. In practice, the time constraints doesn't allow to remove the entire pipe bundle running over the manhole to perform a small job inside a confined space. Therefore, employees are forced to carry out the job with minimum access space, which is most likely to lead to injuries. As such, unless these factors are considered during the design phase of a vessel, employees attending to repair and maintenance throughout its whole lifecycle have to face an undue risk.

6. CONFINED SPACE HAZARDS

According to the Department of Labour (Sri Lanka) Statistics, 83% of onboard fatal accidents reported during this study period of 15 years in Sri Lankan shipyards are inside confined spaces [12].

6.1 FATALITY DUE TO FIRE INSIDE A CRANE PEDESTAL

6.1(a) Case Study

An employee who had been detailed to carryout welding work inside a crane pedestal of a newly built vessel died due to a fire inside the crane pedestal on 24th April 2014. The 28 years old victim had entered the crane pedestal to perform some welding rectification work at the bottom of the pedestal, which was nearly 4m below the entrance opening. There was only one manhole to the pedestal through which the ventilation duct had also been laid to ventilate the confined space. The victim had removed the ventilation duct in order to get inside the crane pedestal since the size of the standard opening was inadequate to enter while the ventilation duct was in place. This way, he had started work inside without forced ventilation, and after about 30 minutes a sudden fire was noted by his confined space attendee, who had called for emergency service. It was noted that the fire was erupted due to oxygen enrichment inside the crane pedestal and during investigation an oxy-acetylene cutter was found inside where oxygen was leaking.

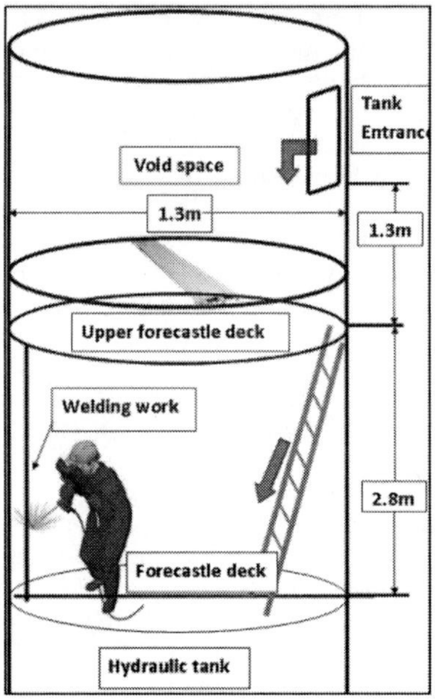

Figure 1: Working arrangement in Crane Pedestal

6.1(b) Cause Of Accident

Due to insufficient space to access the location, the victim has cut down the ventilation to the confined space, which was the main cause for this accident.

6.1(c) Discussion

The fatal fire was due to the oxygen enrichment in the confined space. There was only one manhole for the crane pedestal, which was nearly 4m above the bottom, where the job had to be performed. Due to insufficient space, the victim had removed forced ventilation, which was the major cause for the accident. Considering the HF in construction as well as in maintenance, another manhole somewhere near to the bottom would have prevented the fatal accident. Due to the access limitations, significant time had spent to rescue the victim being another factor for the fatal accident.

6.2 EXPLOSION INSIDE A PUMP ROOM OF A TANKER SHIP

6.2(a) Case Study

An explosion took place inside a pump room of a tanker on 03rd June 2007 causing one dead and 14 injured including the engineer in charge. At the time of the incident, 05 employees attached to one subcontract company had been dismantling a cargo pump, and in the meantime 06 employees had been attending to dismantle a cargo line. The group attending to the pump had used a gas cutter to crop a corroded bolt. At the point of time an explosive gas was trapped inside the cargo pipeline, which was trying to be dismantled, and it had escaped to the pump room and caused the explosion. The fire existed nearly for 45 minutes. One employee was dead due to the explosion and the fireman on duty inside the pump room managed to rescue 13 others. However, 03 employees including the fireman got serious burn injuries, which kept them nearly four months away from employment.

6.2(b) CAUSE OF ACCIDENT:

The explosive hydrocarbon mixture released from the cargo line that had been trapped inside the pump and caused the explosion.

6.2(c) Discussion

The tanker ship where the explosion took place was 228m in length and maximum draft of 14m with more than 67,000 in Deadweight Tonnage. It had been difficult to maintain continuous and effective fresh air supply to the pump room bottom, as it was 14m below the pump room entrance. Hence, hydrocarbon vapour with higher specific gravity had higher chances to trap inside and create gas pockets and cause explosions. Pump room operations are always risky as there is no emergency exit from the bottom unless an opening between pump room and engine room has been temporarily made by the shipyard before their repair jobs. However, cutting this opening is itself a risk. Some situations are faced where the job involvement inside the pump room is less than that of making an opening and refitting. In such a situation, ship owners opt not to provide this emergency exit. As such, pump room in a tanker is another confined area where serious attention to be given in design as far as HF is concerned.

6.3 FATALITY DUE TO ELECTROCUTION INSIDE A SEA CHEST

6.3(a) Case Study

A 28 years old welder with 02 years' experience died due to electrocution while he was performing a welding work on 10th March 2010. The job assigned to the victim was to fabricate, fit-up and weld the compressed air pipe for the port and starboard sides' sea chests of a newly constructed vessel. There were 04 persons allocated to the job including one as a confined space attendee. The deceased was welding the compressed air pipe inside the starboard sides' sea chest and he was standing on the scaffolding located at the dock bottom. The deceased's assistant was standing below the starboard side's sea chest at the dock bottom providing assistance to the welder. Then the deceased's co-worker noticed that their colleague was sliding downward with shaking legs through the gap between ship hull and the scaffolding. Then his confined space attendee tried to help the victim by holding the collar of the overall and he too felt the electric shock. Then one of his co-workers ran and turned-off the welding plant located at the port side of the dock bottom. The victim was rushed to the hospital but at the time of admission he was pronounced dead.

6.3(b) Cause of Accident

At the post mortem it was revealed that the cause of death was electrocution.

6.3(c) Discussion

According to the statements of witnesses, at the time of fatal accident the employee was sweating profusely. His overall was wet due to sweating, which has caused reducing of body resistance. Due to the insufficient space inside, it was difficult to ventilate the sea chest and he had been sweating because of the welding heat. As the space was too small, the welding rod has connected to the welding holder and made contact with the right side of his chest area. Though he had been wearing proper safety shoes, investigation revealed that the path of electricity was completed through his body as he has leaned against the opposite side of the sea chest due to lack of space available. These kinds of operations are not limited to the construction period of these vessels, but during its' life cycle people have to go inside for some other operations, especially for welding zinc anodes. Therefore, proper HF consideration is needed in designing confined spaces such as sea chests.

6.4 ELECTROCUTION INSIDE A STOOL SPACE IN A BULK CARRIER

6.4(a) Case Study

A 28 years' old subcontract employee was electrocuted on 13th August 2005 while he was welding inside a stool space between number one and two holds of a bulk carrier. The particular confined space was having only a single access in the port side corner while the job assigned was at the starboard corner, to complete a welding work close to the sounding pipe of number 01 (port) double bottom tank. While the employee was welding, the above mentioned sounding pipe, which was adjacent to the welding work, suddenly burst and the stool space was flooded with sea water electrocuting the welder. It was revealed that the sounding pipe was badly corroded and damaged prior to the incident. Therefore, someone had wounded the damaged area of the pipe with adhesive tape to prevent water leak. Due to excessive heat generation of welding and gravity pressure of water inside the sounding pipe, sea water had discharged from a damaged area of the pipe and flashed to the welder who was carrying out a welding work and he was electrocuted and died at the accident site itself.

6.4(b) Cause of Accident

Electrocution due to splash of sea water, while performing a welding work was the reason for this accident.

6.4(c) Discussion

Congestion and complicated arrangements in stool space was one major factor for this accident. The welder has to enter from one end of the stool space and walk almost full width of 32m of the ship to reach to his job location. Due to this fact the confined space attendee has a very remote chance of hearing his call and that long distance has reduced the visual acuity as well.

So far, four fatal accidents reported inside confined spaces, during the period between year 2000 to 2015 in Sri Lankan ship repairing and ship building industry, have been discussed. In addition to above accidents, several incidents and near misses were also reported in similar working environments. Insufficient space inside confined spaces is one major challenge while limited access opening is the second critical factor. Further, many tanks, cofferdams and sumps can be found in ships with only one manhole, through which air circulation and forced ventilation are not possible as it disturbs the access way. As such, during the construction space as well as its' operation and repairs workers will have to enter and perform various repair and maintenance work under serious risks for their lives. Therefore, HF consideration in designing confined spaces of ships is required seriously in view of safety of people engaging in various activities during ship repairing and shipbuilding activities.

7. LIFECYCLE APPROACH TO SHIP DESIGN

The above topic has been included in the Safety Guidance for Naval Architects[13]. However it is worthwhile to investigate to what extent HF consideration has been given to each phase of a ship's life cycle. In the ISO 14001:2015 edition, it emphasizes about the Life Cycle Perspective (LCP) of Products or Services with respect to the environmental aspects [14]. Significant attention has been now given and also researches have been carried out with respect to the LCP of various products and services with an environmental perspective. However, so far the attention to analyze the LCP with respect to the safety of the product or service is not yet significant. According to Chatzinikolaou, Ventikos, Bilgili, & Celebi, 2016 [15], a ships' life cycle can be divided into four important phases as far as emission is concerned; these being (1) Shipbuilding, (2)

Operation (3) Maintenance, and (4) Dismantling/Recycling. These 04 stages could provide a suitable platform to assess the Life Cycle Safety Perspective of ships. That is, HF consideration in each phase of the ship's life cycle needs to be separately considered with respect to the context of each situation. As of now, the ship design is more or less confined to the requirements of customer and classification societies in order to get the certifications for seaworthiness. Some researchers have been done to identify the operational requirement and its HF interaction. But the attention drawn on the HF in the other 03 stages, that is Construction, Repair/Maintenance and Dismantling/Recycling is not up to satisfactory level. The above case studies convey clear evidence for the lack of consideration for these aspects. Therefore, HF aspects of a ship should not be confined to one stage, mostly to operational process; instead, it should cover the entire life cycle of the ship.

8. THE MISSING LINK

When compared to other design and construction industries, the number of steps between ship design to construction as well as the number of parties getting involved is considerably high. Generally ship designer, ship builder, ship owner, classification societies and sometimes even the charterer who is expecting to charter the vessel involves in this process. So, as the process is quite complicated, this lengthy and complicated process keeps designers away from Shipbuilder which is one of the key issues in this industry. Therefore the issues encountered by shipbuilders during the construction period are not communicated to the designers. There are numerous HF issues in the construction phase which employees are prone to serious risks, but the ignorance or oversight of HF during design stage may not be referred to the original designer. When it comes to the ship repair, the situation is worse. Ship repair yards hardly communicate with ship designers about the HF issues they are facing in the repair stages. For that reason the designers are having a very remote chance to get their designs corrected or improved through a proper feedback process. After the shipyard's accident investigations, some fault of their employees would be identified as a root cause of the accident and may not look at the big picture of the accident. However, human contribution to accidents are summarized by Reason[16] as shown in Figure 2, suggests that accidents are not just a result of an action of an individual, instead it is a failure of a sequence of processes and decisions. So then the shipyard worker is the one who triggers the accident or breach the last level of defence, but the environment that causes the triggering of the accident could be due to a lapse in the initial design. In Figure 2, the term "(at the design stage)" was added to the first line of defence to customize the Reason's[15] theory to this discussion.

As highlighted in the "Incorporate good practice into Ship design process" [17], the future ship designers should meet end users to identify the real challenges in shipbuilding and ship repair activities in order to provide a sound ship with appropriate HF consideration. Professor Kuo[5] also highlights with some examples that there are good reasons to integrate Human factors into every phase of a product's life cycle. Therefore establishing of this missing link between ship designers and shipyards (shipbuilders and ship repairers) is of utmost importance in promoting better safety standards in shipyards.

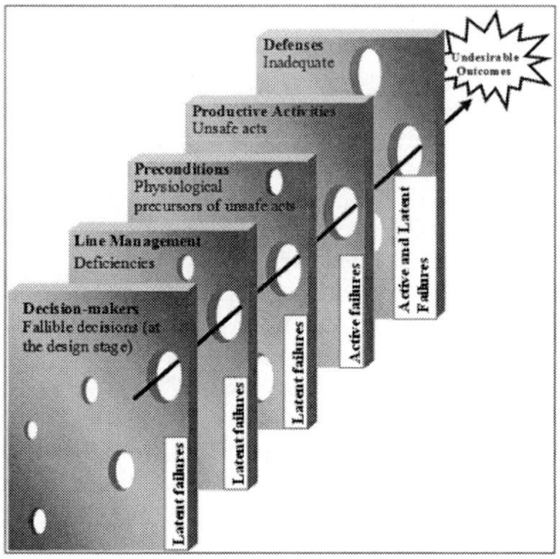

Figure 2: Swiss Cheese Model of Reason[15]

9. CONCLUSIONS

A ship itself is a very complicated design since many systems have to be accommodated in a limited space. On the other hand, there are so many interested parties for a ship including owner, charterer, shipbuilder, classification societies and flag state of the ship, and there are various demands and rules or guidelines, which should be considered by the designer during the design stage. Ships are always huge financial investments and hence investors expect a better return for their investments. As such, the priority for safety can be oversight by commercial pressure. Therefore, repercussions, which may arise in operation, running or maintenance of a ship as a result of any omission during the design stage, will be hardly affordable by the respected parties. Hence, it is of utmost important to take proper account of the interaction between the ship and people who use it.

This paper highlights some of the shipyard accidents, which could have been prevented if proper HF consideration had been given at the design stage. Out of many factors discussed, the competency gap between qualified seafarers and outsourced employees in shipyard activities is an important factor. Further, when considering the repairing stage, shipyards are having very limited freedom to make temporary fencing and guards as far as time and fund allocation is concerned, and therefore, unprotected openings, including manholes and ladder openings are another factor to be taken into consideration at the design stage.

Ergonomic consideration in designing manholes and access ladders is also an important area as it causes frequent sprains and accidents for ship repairing and ship building workers. Thereby, providing of uninterrupted access with proper handholds at accessible heights are key areas for consideration. Working in confined spaces of any work site is a serious challenge due to limitations of ventilation, illumination, rescue operation, space restriction for movements and number of accesses. Therefore, it should be highlighted that providing at least two access points may significantly reduce the risk of employees attending to confined space work.

The evidence of accidents illustrates the importance of integrating lifecycle approach to ship designing, concerning the significance of the HF. It is clear that some of the aspects such as construction and maintenance of ships are not properly captured at the initial design stage and needs to be recapped?. In connection to the same issue establishing a sound feedback loop by ship designers are very much essential. Hence, there should be a communication protocol from shipbuilding and ship repairing yards to designers on fatal / serious accidents or near misses occurred, which could have been eliminated by a simple design change or alteration. Such feedback would definitely help the designer to improve the HF consideration in the said area and improve safety as well as efficiency of repair or construction work.
Further studies on this area with more shipyard accident data would help ship designers to improve the human factor in ship designing in greater ways. Although this study was focused mainly on shipbuilding and ship repairing aspects only, when it comes to the lifecycle of a ship, consideration for ship recycling too needs serious attention since the average fatalities in a ship breaking yard has been 14 per year over the last 30 years as per the records. However, the unofficial figure of fatalities per year is around 67[18].

10. ACKNOWLEDGEMENTS

A formal declaration of acknowledgement is arduously sufficient to avow my gratitude towards the Sasakawa Peace Foundation and specially to its chairman

Dr. Yohei Sasakawa who afforded a considerable sphere to study at the World Maritime University, which verily upgraded my academic career.

Furthermore, I am grateful to the former MD/CEO of Colombo Dockyard PLC, Mr. Mangala Yapa, who played an exemplary role in my professional career and encouraged me to further broaden my knowledge.

Withal, I respect my contemporary employer, Colombo Dockyard PLC and its Management which gave all assistance to develop my career in the Maritime Industry. In the meantime Mr.D.VAbesinghe (Chief Operating Officer) of the Dockyard too should be kindly mentioned for encouraging me to publish this paper.

Ultimately, I am indebted to my amiable wife for her genuine encouragement, support and quiet patience during my academic engagements. Likewise, my beloved two sons and my parents for their untiring love and support for my academic career.

11. REFERENCES

1. Health and Safety in Shipbuilding & Ship repairing, ILO Code of practice, International Labour Office, Geneva, 1974

2. Beck, U. (1992). Risk society: Towards a new modernity. (2008 ed.). London: Sage Publications

3. Perrow, C. (1984), *Normal accidents: Living with high-risk technologies*. Basic books, Inc. Publishers, New York

4. Colombo Dockyard PLC. (2009)- Accident Analysis report

5. Kuo, C. (2007). *Safety Management and its Maritime application*. United Kingdom: The Nautical Institute. (Kuo, 2007)

6. Bea, R. University of California, Department of Civil & Envionamental Engineering. 2010). *Failures of the deepwater horizon semi-submersible drilling unit.*

7. Reason, J. (1993) Managing the Management Risk – new approaches to Organizational Safety. in: Wilpert, B. and Qvale, T. (Ed s). *Reliability and Safety in Hazardous work systems: Approaches to Analysis and Design.* LEA Hove

8. IKEI, 2009, Comprehensive sectoral analysis of emerging competences and economic activities in the European Union: Building and repairing of ships and boats sector

9. Hart, Pieter 't, and Dick Schotte. "HR-Forshungsstudie (2008): Demographic Change & Skills Requirements in the European Shipbuilding& Ship Repair Industry." *Google Books*. European Commission, n.d. Web. 25 Aug. 2016

10. Yilmaz, A. I., Yilmaz, F., & Celebi, U. (2015). Analysis of Shipyard Accidents in Turkey. *British Journal of Applied Science &*

Technology BJAST, 5(5), 472-481. doi:10.9734/bjast/2015/14126

11. Annual report, Colombo Dockyard PLC, 2014

12. Accident Analysis 2000 to 2015, Colombo Dockyard plc

13. The Royal Institute of Naval Architects, "Safety Guidance for Naval Architects", Second Edition, 2010

14. ISO 14001:2015

15. Chatzinikolaou, S., Ventikos, N., Bilgili, L., & Celebi, U. B. (2016). Ship life cycle greenhouse gas emissions. In *Energy, Transportation and Global Warming* (pp. 883–895)

16. Reason, J., Managing the Management Risk: New approches to organizational safety, University of Manchester

17. Abeysiriwardhane, A., Lutzhoft, M., Petersen, E.S., and Enshaei, H. (2016). *Incorporate good practice into ship design process; Future ship designers meet end users.* Paper presented at the *ERGOSHIP 2016: Shaping Shipping for people,* 6-7 April 2016, Melbourne

18. Sahu, G. (2014), Workers of Alang-Sosiya; A Survey of Working Conditions in a Ship-Breaking Yard.

12. AUTHORS BIOGRAPHY

Prasad Manjula Hettiarachchie (FIIRSM, RSP) is the Management Representative/Manager (Compliance) of Colombo Dockyard PLC, a well established and a leading Shipbuilding and Ship repairing facility in Sri Lanka.

Manjula launched his career in Colombo Dockyard PLC as a *Safety Executive.* That eventually led to a 16 years partnership with his company and position himself as the Manager (Compliance)/Management Representative. It is in this capacity that he performs as an independent and objective body that reviews and evaluates compliance issues/concerns within the organization giving special attention to Safety, Health, Environment and Quality aspects. His position attests the Management and employees are in compliance with the applicable legal and other requirements applicable to the organization. Withal, that behaviour in the organization meets the company's Standards of Conduct. During his professional career, he has been practically involved in developing, implementing and maintaining an Occupational Health and Safety Management System in the company. He has been attending to give safety clearances, safety assessments, audits and accident investigations throughout his career in Colombo Dockyard PLC.

Manjula acquired his M.Sc in Maritime Safety & Environmental Administration from the World Maritime University, Sweden in 2012. In his ceaseless route he obtained his Master of Business Administration degree from the University of Colombo in 2005.

Manjula has qualified from the Engineering Council Examination (UK) in 2005 satisfying the academic requirements for registration in the UK as a Charted Engineer.

OPTIMIZATION OF THE ERGONOMISTS INTERVENTION INTO A SHIP DESIGN PROCESS

C Stalla, J Flohic, DCNS, France

SUMMARY

Mainly through major programs of surface ships and submarines for the French Navy, DCNS has experienced Human Factors and Ergonomics (HFE) integration into the design process on a large-scale. A broad spectrum of approaches could be experienced by the DCNS HF team in the way to promote the HFE point of view in the design process such as: normative approach, missions and tasks analysis, use of simulators, management of a broad participation of future users in the design process at both definition and evaluation stages. The experience is mainly about FREMM frigates project and BARRACUDA submarines project but also includes smaller projects. It goes from very early stages of the design to commissioning phases.

This paper presents the main lessons learned by the DCNS HF team in the field of HF integration over the past decade. These lessons result in a number of suggestions about the best possible use of the ergonomics discipline in naval projects: tools and methods but also ergonomists themselves. In the perspective of cost effectiveness of HF interventions some tracks are proposed at several stages including: reasonable contractual commitments and management of the relations between shipowner and ship vendor in the field of HFE, collaboration between disciplines, position of ergonomists in the innovation process at whole ship level and at equipment level.

Keywords: Human Factors Integration, ergonomics, crew, naval ships, Human Centered Design.

DCNS

DCNS group is a leader in naval ships design. It is positioned as prime contractor and integrator for whole warships. It designs, manufactures and provides for the most innovative surface naval ships and submarines on the market along with the associated critical systems.
DCNS is also positioned as a major actor in marine renewable energies.

This paper focuses on Human Factors (HF) in naval ship process design.

NOMENCLATURE

ANEP: Allied Naval Engineering Publication
CAD: Computer Aided design
CCD: Crew-Centered Design
CEN: Comité Européen de Normalisation
CMS: Combat Management System
HCD: Human-centered design
HF: Human Factors
HFE: human Factors and Ergonomics
HMI: Human-Machine Interface
HSI: Human System Integration
IBS: Integrated Bridge System
IPMS: Integrated Platform Management System
ISO: International Organization for Standardization
NATO: North Atlantic Treaty organization
Q&A: Qualification & Acceptance
R&D: Research & Development
RINA: Royal Institution of Naval Architects
UCD: User(s)-Centered design
UK: United Kingdom

1. HUMAN FACTORS & ERGONOMICS STUDIES IN SHIP DESIGN PROCESS AT DCNS: OVERVIEW

1.1 HISTORY

The intervention of qualified ergonomists in DCNS naval projects raised in the decade 1990. This historical collective experience of the DCNS Human Factors team has been deeply enriched through two major programs of the French navy launched in the decade 2000: FREMM frigates (Fig. 1) and BARRACUDA submarines (Fig. 2). A broad spectrum of tools and methods could be experienced by the DCNS HF team in the way to push the HFE point of view in the design process. It goes from very early stages of the design to commissioning phases, from stern to bow and from mechanical systems to software systems, within huge industrial organizations.

Figure 1: FREMM frigate from DCNS

Figure 2: BARRACUDA submarine from DCNS

1.2 SCOPE OF THE DCNS HF TEAM

For DCNS, Human Factors and Ergonomics in ship design activities refer to a global approach that includes:

- Manning plan (crew) & work organization according to the ship-owner policy regarding the human resource and consistently with the design (missions, functions, level of performances)
- All aspects of ergonomics (physical and cognitive ergonomics).

The DCNS HF team includes several ergonomists disseminated in DCNS industrial sites.

The core activity of these ergonomists consists in being involved in industrial design teams so within a contractual environment. Ergonomist(s) are then responsible for the integration of HFE in the project. "Ship design process" is then extended here to pre-contract and contractual phases including Qualification & Acceptance (Q&A) activities.

In addition, the DCNS HF team carries out others noticeable activities that strengthen the effectiveness in the core activity:

- Contribution to R&D studies. This applies to key innovations according to their HFE dimension;
- R&D studies aimed at improving the set of tools and methods related to HFE;
- Education courses about HFE for naval engineers and technicians.

1.3 USUAL SERVICES PROVIDED BY THE DCNS HF TEAM WITHIN THE SHIP DESIGN PROCESS

Efficiency at work onboard is the most common motivation for requesting the intervention of the DCNS HF team. The request can come from the inside of the company (chief of project, technical designer, marketing,..) or from a client. The typical interventions aim at user-friendly products and so make a contribution to safety. Regarding safety, it is underlined that the compliance to Occupational Health and Safety regulations is not in the perimeter of the HF team. In the organizational structure of DCNS engineering staff there are dedicated teams of specialists in charge of each of the important subjects such as Safety and Occupational Health, Maintainability, Nuclear Safety, etc. Thus according to the topic the HF point of view of ergonomists can be requested from specialists of others disciplines and vice-versa, as necessary. At a whole warship or submarine level, design activities are, of course, multidisciplinary.

Considering the product (surface ship or submarine) the field of intervention is very open. The potential areas of influence include the overall architecture down to any component of the ship (Fig. 3):

- General Arrangement;
- Spaces lay-out;
- Consoles and workstations including the arrangement of controls and displays;
- Human-Machine Interface (HMI); and
- Features of the (sub-)systems: utility & usability, levels of automation.

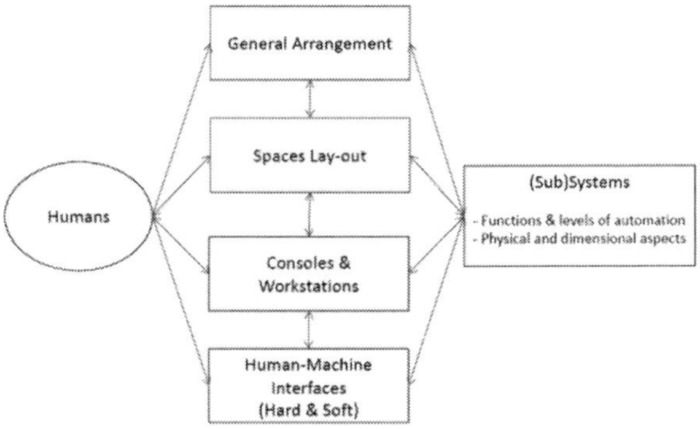

Figure 3: Typical pattern for a general ergonomic intervention within a ship design process

All these elements will be analyzed through a specific point of view which is about "how to use it". This balances the thought pattern from engineers and technicians which is mainly about "how it runs". At the core of this singular point of view are always the users, keeping in mind that the product is a socio-technical complex system, with many interactions.

Within the ship design process the DCNS HF team uses to perform:

- "Stakes analysis" of the project from a HF point of view. This is aimed at highlighting the key matters of a project from a human centered perspective considering not only the product in itself but also

the possible impacts of the industrial organization or the purchase policy;
- Management of HF integration in the project. This includes contractual aspects i.e. technical requirements on the product but also HF activities and the associated methodologies, planification of HF activities consistently with the design and purchase process;
- Missions and tasks analysis in order to produce useful data for designers (from experience based on visits on-board alongside, shipping periods, interviews with internal operational experts (periods of or exchanges with users' representatives within official working groups);
- Experimentations, assessments of systems and/or arrangements. For arrangement the range of tools goes from symbolic representation to virtual reality (Fig. 4 and 8), or scale 1 mock-up (fig. 5), according to the level phase of the project. Specific tools such as digital manikin can be proposed. For the validation of innovative control & monitoring organizations and associated software the use of simulators or prototypes can be considered;
- Leadership of working groups including future user's representatives, according to the needs for feedbacks from users and according to evaluations or validations of the solutions suggested by designers or vendors;
- Management of the contractual technical requirements assigned in the HF perimeter, including ergonomics standards according to their contractual status. This goes from contractual discussion to acceptance and qualification phase as necessary; and
- Assistance for technical designers ("hot line") as far the pro-active ergonomic intervention has not reached them.

Figure 4: virtual reality room

Figure 5: validation of shape and dimensions of a console with user's representatives using a scale 1 adjustable mock-up

2. GOALS AND CONTEXT

This article wishes to be a contribution to the development of user(s)-centered design (UCD) approach. UCD such as named her refers to Human-Centered design (HCD) or Crew-Centered Design (CCD) according to the sources. This contribution is through the testimony of applying it since years in DCNS projects. The experience includes FREMM frigates and BARRACUDA submarines projects, but also smaller projects.

Military ships and submarines are among the most complex products. Following the extension of automation, internal complexity of sub-systems combined with numerous interfaces between them lead to extremely tricky but exciting design activities: how to make that simple and easy-to-use, even in case of failure?

1st of class AQUITAINE frigate is now sea proven for several years and it seems that despite the level of sophistication, the challenge has been globally succeeded with some fields of high level of satisfaction from the users such as in the field of control & monitoring of platform systems.

1st of class SUFFREN attack submarine is not launched yet. However confidence is now permitted as the activity of representative users could be observed on software integration tools and more recently on training simulators.

Beyond the testimony about applying UCD principles, the article also wishes to develop the awareness about ergonomists as the belief is that the implementation of UCD would better involve their specific skills. Keeping in mind that ship design process are inevitably under costs and time constraints, some suggestions are made about how to maximize the benefits of having ergonomist(s) in a design team, within a collective strategy for Human System integration in the project.

UCD is long to disseminate but it is not the latest fashion. Already as from 2002, R.V. POMEROY & B.M SHERWOOD JONES [Ref. 1] warned the naval community about the limits of a prescriptive approach intended at HF integration in ship design process :"*For situations where there is any significant change in the manning, technology or type of operation, there are major difficulties with producing detailed descriptive design specifications*". They recommended a step forward to goal-setting approaches based on user-centered methodologies. The concrete application of user-centered approaches within DCNS naval projects especially during the last 10 years confirms how relevant this statement is. Analyzing the future work is a fruitful way to raise key factors. Principles associated with usability are quite easy to understand but they are rather qualitative so transforming them into specific measurable requirements from the beginning as expected by the technical designers is a tricky exercise. As the human activity is influenced by the technical solution and as the technical solutions is expected to be consistent with the activity, an iterative process is necessary to converge. But despite this theory is generally agreed by specialists of design process of complex systems, the following statement, made nearly 15 years ago during the 2002 RINA Conference seems to be still an extent today: "*Suppliers have evolved a skill base, culture and costing strategy based on straightforward compliance with specific requirements*". In a contractual environment, whatever could not be specified at contract signing may be rejected during the design process. And qualitative principles such as "easy to use" are unwelcome by technical designers. They are suspected to be too subjective so that they generate risks of future costly negotiations with the client. Considerable efforts are then required in order to convince that UCD will strengthen not only the design in itself but also the relationship between the stakeholders, and so contribute to a peaceful acceptance phase.

Another important element of the context for any ergonomist embedded in a design team is time and budget constraints. The intervention of ergonomists is not for free. This strengthens the expectation for straightforward solutions and basic stabilized criteria rather than the holistic mind-set associated with UCD.

Vocation and perseverance are mandatory for anyone involved in the integration of HFE in the ship or submarine design process. It is in respect of this concrete context that the following opinions and suggestions are made. The objective is not to claim full solutions. It is humbly aimed at stimulating a debate about the possible ways to concretely and at reasonable costs:

- Provide "ergonomic products" (safe, easy to use and efficient) to the ship-owners and so seafarers;
- Develop as necessary the involvement of qualified ergonomists within the naval community; and
- Optimize the industrial investments in qualified ergonomists.

The subject raises a number of questions about:

- How to contract about HFE in a concrete ship design programme which means under time and cost constraints;
- The dissemination of knowledge related to HFE among technical designers; and
- The interactions between ergonomists and the various disciplines involved in naval projects.

These questions will be treated along the following structure:

- § 3 will underline the clients' influence and responsibility in the current context;
- § 4 aims at sharing the normative background and then introduce the possible contractual implications;
- § 5 aims at showing that it is possible to implement UCD in a contractual environment. This is achieved through factual description of DCNS practices in catching and processing the users 'needs;
- § 6 will underline the need for strengthening the dissemination of HFE knowledge in the maritime community but also the need for qualified ergonomists in addition; and
- § 7 will draw subsequential suggestions aimed at optimizing both the contractual structure and the industrial organizational structure in the purpose of global industrial collective efficiency regarding Human System Integration (HSI).

3. CLIENTS' INFLUENCE ON INDUSTRIAL EFFORTS REGARDING HFE

In DCNS, the HF team has that chance that the main customer, which is the French Ministry of Defence, has an incentive policy about HFE. According to the project, the ergonomic interventions are formalized within the contractual documentation. This give legitimacy to the ergonomists and it enables the development of the discipline within the ship design activities.

In the military domain, some governments have developed their own literature. But regarding the worldwide globalization, homogenization of practices would help in contracting about HFE in ship design contracts. Although some aspects of HFE are covered by the Rules of classification societies (level of noise and vibrations...), there is no kind of certification from classification societies or any other relevant entity that would explicitly cover the integration of HFE at whole ship level. So the level of efforts regarding HFE still strongly depends upon the customer's request.

R.V. POMEROY & B.M SHERWOOD JONES [Ref. 1] underlined that moving to goal-setting approaches instead of the traditional prescriptive approach will be effective upon the condition of "*an informed purchaser and an intelligent supplier*". In this perspective there is a mutual interest that:

- The clients are well informed and have sufficient guarantee that all desirable efforts regarding HFE will be initiated as necessary by the design company but also by vendors of systems;
- The ergonomist(s) embedded within the industrial design team is (are) in the best possible conditions to maximize the gain from his (their) intervention regarding the financial budget and the collective skills of the design team regarding HFE.

Thus, in the current context of no mandatory process regarding Human System Integration (HSI), a pre-requisite for the optimization of the ergonomic intervention is about the contract. The very first activity is then to define as clearly as possible the optimal scope of and the optimal level of HFE efforts. This is the result of the so-called "stakes analysis" practiced by the DCNS HF team.

Once a shared comprehension of the stakes is reached, in a general manner the difficulty is to find the operating point between vague orientations and too low-level requirements. Specifying a methodological standard is not enough and would increase the risk of divergences later on about the perimeter and the depth of the investigations. By contrast, focusing on some ship systems may inhibit the open-minded approach.

The client has a key role in the optimization of the industrial efforts regarding HFE. As no unified mandatory process exists, it is essential that he assumes his part of responsibility. Specifying nothing or specifying the moon will lead to non-optimized HFE efforts. Awareness about the background in ergonomic standards is then a basic but crucial point.

4. CONTRACTUAL PROBLEMATICS RELATED TO STANDARDS IN HFE

Normative landscape in ergonomics

In DCNS environment the standards related to human factors and ergonomics can be classified according different kinds of logic:

- Material (including generic Human capabilities and weakness) / Methodological;
- National / International;
- Military (restricted or not) / Civilian; and
- Maritime / Generic.

For instance the background of DCNS ergonomists includes (non-exhaustive list):

- "Historical" NATO military guidelines published in early 1990's : Guidelines for shipboard habitability requirements for combatant surface ships (ANEP-24), Guidelines for environmental factors in NATO surface ships, Ergonomic data shipboard space design into surface ships (ANEP-26);
- MIL-STD-1472 Human Engineering (US standard);
- DEF-STAN 00-250 Human Factors for Designers of Systems (UK standard);
- ASTM F1166-07 Standard Practice for Human Engineering Design for Marine Systems, Equipment, and Facilities;
- ASTM F1337 Standard Practice for Human Integration Program Requirements for Ships and Marine Systems, Equipment and Facilities;
- ABS Guidance Notes on the application of ergonomics to marine systems; and
- ISO 9241 (formerly ISO 13 407) related to ergonomics of human-computer interaction.

In 2004 the 39th congress of the *Société d'ergonomie de langue française* focused on this specific matter of standards in ergonomics. It was already underlined the expansion of ergonomics standards. It could be found around 150 ISO and CEN standards related to ergonomics being around 3000 pages [Ref. 2]. You can bet that the number has grown year after year up to now.

Considering the number of standards and a generally poor educational knowledge about ergonomics among naval engineers and technicians (almost about methodological aspects), it seems useful to underline the problematics related to what we have named "material standards" and "methodological standards".

"Material standards" will refer to standards or parts of standards that address technical requirements on the product or on its components (widths of passageways, size of characters according to the distance of the operator…). This category of standards might be used by ergonomists but also by technical designers themselves according to their ability, as input data and criteria. MIL-STD-1472 Human Engineering (US standard) is probably the most famous one but is not specific to ship. ASTM F1166-07 has about the same philosophy and presents the advantage to be aimed at ships. This kind of standard is quite familiar for engineers and technicians as the content is rather at detailed specification level.

"Methodological standards" refers to standards or parts of standards that address process and methodologies. They are not well disseminated among technical designers, project managers and corporate decision-makers. This kind of standards is aimed at ergonomists but not only. It is also aimed at project managers as it is their responsibility to involve ergonomists or not in their project. The most representative standard in this category today is probably ISO 9241-210 Human-centered design for interactive systems.

Risk of overloading the design process with too many low level requirements from "Material standards"

When entering a design phase the usual reflex of designers consists in asking for the requirements to comply with (thinking of Material requirements).

In the name of Quality assurance some people would expect that any item of a standard specified in a contract is allocated unitarily through the design process. According to such a strategy, ergonomics standards would have to be divided into elementary items, sentence by sentence as necessary. Any elementary item would then be labeled with a unique reference and then allocated according to a functional or product breakdown structure or according to other logic such as topography. This kind of approach is based on the wish for a total control and the belief that the sum of elementary compliances will give the global compliance. This is typical from the prescriptive approach. In this approach, standards are considered as a constraint to comply with in a very strict manner. According to the Quality assurance process detailed compliance matrix are then expected at commissioning phase including the production of evidences for any elementary item. But such a policy can lead to hundreds of items (for instance: MIL-STD-1472 is nearly 400 hundreds pages long!) allocated to hundreds of technical elements (surface ships and submarines are among the most complex products). At whole ship level this generates a risk for overloading the design process and almost the validation process with a high number of couple requirement/evidence. In fact it is simply unrealistic at whole ship level under the constraint of reasonable costs. From experience some people that will claim first for labeled detailed ergonomics requirements will in a second time, once confronted to reality of ergonomics standards, claim for a limited number of inputs!

The whole technical solution may not be available in the "Material standards"

Moreover, the contents of ergonomic "material standards" are more or less generic or even vague according to the topic. A basic example is about the anthropometric data. Several standards propose default values for anthropometric data taken from databases. However, it is recommended to fit the anthropometric data according to a projection regarding the population of future users. So when designing a ship that will still be in service 40 years later, one should consider anthropometric that are not from databases. Designing a ship for people nearly 2 meters tall may lead to some adjustments compared with current values stated in the standards. The limits are not only regarding dimensional aspects. For instance, C. CHAUVIN & G. LE BOUAR underlined [Ref. 3] that ISO 8468 about Ship's bridge layout, does not give the designer a complete solution especially for specific ships. In order to properly arrange the bridge and the devices on the console it is necessary to define the work organization. This could be difficult to accept for non-informed people as far this standard is specific to the maritime domain and deals with a unique room. Indeed, the experience on DCNS project is fully in line with these authors, not only considering the ISO 8468 about bridges but as a general inevitable limit of "material standards". In the literature there are many cases of fully compliant design regarding this kind of standard that are very far from user-friendliness. This sad truth is not specific to the maritime domain but this also applies to it. That does not mean material-oriented standards are worthless. The point is just about being aware of their limits. Combined with the risk of overloading, this should lead both the client and the industry to weigh carefully the range and almost the status of ergonomic "material standards". This should encourage a discussion about how to withstand the problem, which means discuss about a relevant methodology to overcome it.

Methodological standards: users-centered design

Fig. 6 is adapted from ISO 9241-210 and describes the typical methodology promoted by the community of ergonomists.

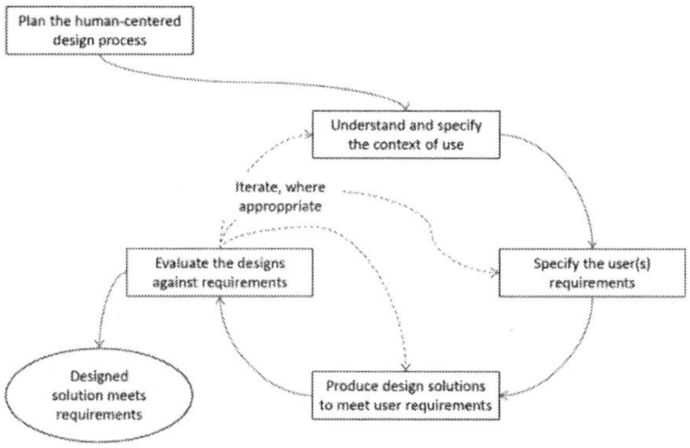

Figure 6: Human centered design generic process (adapted from ISO 9241-210:2010)

In this approach, the user's needs are to be caught through a holistic approach. This is not exclusive from using "material standard" It is just that only the relevant data will be taken from the most relevant "material standards" and possibly adjusted according to the context, as necessary. Here "Material standards" can be considered as a resource to be used mainly while producing solutions rather than a requirement in itself.

It is to be noticed that human-centered approach is now requested in planes design process to obtain the certification of the product from the authorities. According to the stakes, some ships to be designed may also deserve tailored UCD process.

5. CATCHING AND PROCESSING USERS' NEEDS

The principles of UCD were fully experimented across FREMM frigates and BARRACUDA submarines projects. In each of these projects, a HF integration Plan was established in order to formalize the scope of work, the general process and the milestones consistently with the "classical" design process at whole ship and at technical systems levels. Regarding the ISO 9241-210 this corresponds to the planification step. Due to the size and length of such projects, the HF Plan is at milestones and principles level. The short-term and medium term planning of the meetings with user's representatives was discussed and updated as necessary between industry and client HF managers,

The scope included key subjects related to the effort in optimizing the crew and the overall operational performance. Beyond that the aim was to limit the risks of late and costly change.

For instance, the HF intervention on the mooring decks on FREMM frigates started while technical studies were just initiated. Dimensioning scenarios were analyzed with relevant users 'representatives. The working Group detailed as far as necessary the individual tasks but also the variability factors. The collective dimension was also addressed including the interactions within a mooring deck and the interactions between the bridge, each of the mooring decks and the dock operators. In order to preserve an open mind-set, very symbolic supports were used such as 2D- basic scheme and magnets to represent the operators (Fig. 7). This allowed to easily share a comprehension of the activity, including the movements of the operators at each step of the scenarios. From the analysis it appeared that it was possible under certain conditions to mutualize one of the operators compared with the usual practice. The activity-centered discussion raised several recommendations at arrangement level but also at systems level. For instance it was identified the impact of the shape of the winches, so that this concern could be integrated in the purchase specifications.

Figure 7: Working Group session with representative users

Later on a 2D drawing of the arrangement could be agreed within the working group. Then a 3D preliminary mock-up was used in virtual reality so that representative users were able to project into the future situation (Fig 8). The virtual reality tool allowed an evaluation of the visual impact of stanchions. Whereas some doubts had occurred using basic CAD tools it appeared thanks to virtual reality that the situation was acceptable.

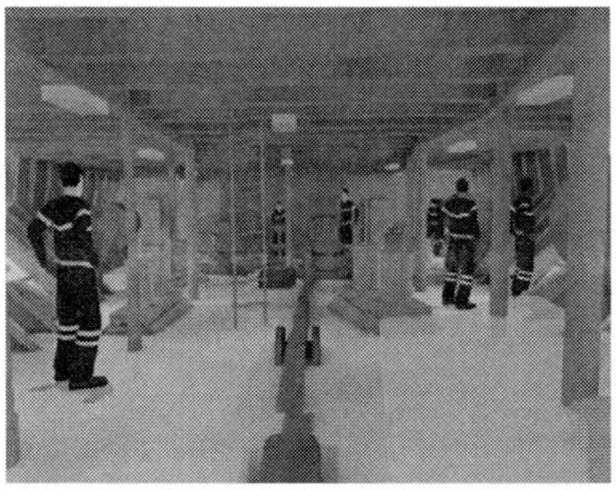

Figure 8: Virtual reality mock-up of a forward mooring deck

Thus, the very detailed arrangement of the mooring decks could be launched without major risk.

According to the stakes, a major part of the HF efforts were about control rooms including the control & monitoring systems: integrated platform management system (IPMS), integrated bridge system (IBS) and combat management system (CMS). Government-owned simulators were used at early stages of the project so that the human activity related to innovative organizations and innovative software functions could be observed.

Some observations were performed jointly by HF specialists from industry and HF specialists from the Government, as a piece of the UCD process.

Between 2006 and 2015, FREMM and BARRACUDA projects together represented around 200 meetings with relevant users' representatives. This includes experimentations in early stages using simulators and validation phases on software integration platforms. Around 4000 contextualized recommendations were issued from the interaction with the users 'representatives. Each request was managed by the DCNS HF team so that for any of them a decision was taken within Industry at relevant level regarding the decision to implement it or not. All of these recommendations were thoroughly managed. The progress was periodically presented to a Steering Committee headed by the Client and tricky topics were then negotiated in the contractual environment at relevant level as necessary. This organizational aspect made that the users 'need had to be conveniently explained and justified. Keeping the Working Groups level rather neutral and objective was probably a key factor of the success in the application of UCD.

6. DON'T FORGET ERGONOMISTS!

Unfortunately the ergonomics discipline suffers from a lack of awareness about the added-value of ergonomists and even more about their methodologies. This is probably due to the fact that ergonomics is rarely or poorly integrated in initial engineering educational programs.

Thus it is usual that project managers, naval engineers and technicians who have never met an ergonomist before have only a vague idea about what they can or cannot expect from an ergonomist. Therefore, most of those who have never experienced a genuine user-centered methodology, leaded by an experienced ergonomist, usually believe that ergonomics is mainly about common sense or applying some basic material criteria, such as the standard height for a table, human vision and so on. They feel like autonomous even though unaware of key concepts and the mind-set regarding ergonomics is usually about prescriptive design at low level of detail.

Some technical designers have the feeling that ergonomics is not only common sense but do not know well how to interact with ergonomists nor when. And once they call the ergonomist for the first time it may be a bit late so that the situation can be painful for them.

From the experience across ship design teams it appears that a few engineers believe (or would appreciate) that ergonomists have a magical power or magical tools. A typical expectation from the ergonomists is about predicting the future occupational workload or behavior. But this is not just like calculating speed, weights or any other physical usual parameter of a ship. The scientific evaluation of cognitive workload requires experimentations with representative users. But on early stages of a project, the descriptions of the hypothetical functions, performances and level of automation of the technical system may be so vague that the prediction will rather be more an initial hypothesis to be refined along the design process. Those designers are on the good way for taking into account the user in their design. But they also need to be advised about what they can and cannot expect from the ergonomist.

Of course the above description of designers' attitude regarding ergonomics is a kind of caricature, but is not mockery. The lack of awareness about the discipline is probably also due to a lack of communication by the ergonomists community. It is ergonomists' responsibility to disseminate the relevant knowledge among technical designers and decision-makers. Hopefully some events such as the RINA Conference about Human factors in Ship Design & Operation give such opportunity and helps in this direction at whole naval community level. The impact of the EU funded project CYCLADES, which was aimed from 2012 to 2015 at developing Crew-Centered Design (CCD) in the maritime community, also raises hope.

But it seems illusory to have any technical designer definitely trained in HFE. R.V. POMEROY & B.M SHERWOOD JONES [Ref. 1] claimed that *"the human element must become part of the mainstream rather than a specialist discipline"*. It is certain that increasing the skills of technical designers in the HFE is necessary. But a one-day or even one-week training will not turn them into HF specialists. The typical educational level of ergonomists is rather at Master or PhD. It is a specialty in itself. Adopting the HF point of view (changing mind from "how it works" to "how it is used" and being able to catch the user's needs) is not so trivial and requires a specific mind-orientation. Moreover, as the activity is not fragmented such as product breakdown structures are, a global approach remains necessary. Architects at whole ship level have already many things to deal with. The more complex and innovative the ship is the most the support from an ergonomist is useful for him.

In DCNS, technical designers (naval architects, specialists of technical systems, CAD technicians...) can attend training sessions about HFE as part of their individual training program. Of course the aim is to disseminate a culture and make them as autonomous as possible. But it is also intended at making them aware of their own individual limits and informing them about the support they can get from qualified ergonomists as necessary. The matter of HFE is so huge at whole ship level that a collective multi-disciplinary strategy will be the most efficient. Qualified ergonomists plus informed technical designers should be the standard practice within design process of innovative ships or ship systems. There is a dynamic and paradoxical dimension in the sense that the more technical designers collaborate with ergonomists, the less they need ergonomists. However, on the base of years of practice, the involvement of ergonomists within industrial design teams, not only as tool-makers but as full actor of the design, will remain relevant for a long period.

Due to the extent of the matter at whole ship or submarine level and due to the duration of naval projects (almost military ones), the investment in ergonomists is inevitably a long-term investment. Many years are necessary to get a practical experience of HFE at whole product level and for each phase of the process. Of course, the financial point of view is to be kept in mind and the investment in qualified ergonomists has to be in accordance with the stakes and the level of innovation (technical or organizational change compared with previous design). The return on investment is quite difficult to quantify. Thus the decision for employing ergonomists within engineering staff requires a strong belief from industrial decision-makers at executive level. But the clients may also have a significant influence on the level of the industrial efforts regarding HFE.

7. SUGGESTIONS REGARDING CONTRACTING AND ACCEPTANCE OF HFE ASPECTS IN SHIP PRORAMMES

Employ ergonomists on their most valuable skills

The experience confirms the added-value of UCD. Therefore ergonomists should be involved in priority in implementing conveniently UCD. From a contractual point of view, the ISO 9241-210 is currently a relevant reference. However no standard will replace a statement of work describing the scope of the intervention. This scope is to be agreed initially on the base of a preliminary stakes analysis at mission level considering the level of technical or organizational innovation. The scope should usually be specified through a breakdown of users 'activities rather than through a product breakdown structure. For instance, analyzing the human activity regarding control & monitoring of the platform is not only related to the Integrated Platform Management System (IPMS). The point of view is different than the thought pattern of technical designers. This maximizes the integrative effect from the users' point of view. The contract and/or a HF integration plan will set the principles for catching and processing users 'needs as well as the associated decision-making process in the contractual environment.

Associated with the statement of work, a piece of the contractual specification should define at least:

- The user's characteristics: anthropometry, language…;
- A list of the "material standards" that will be the default resource as necessary (Cf. § 4); and
- The scope of "transverses goals" in term of work onboard the ship to be designed. Relevant associated contractual documents should detail Qualification and Acceptance aspects (Q&A) aspects. For each goal, the kind of evidence that is expected regarding the goals should be clear.

Impact of the workshare between technical designers and ergonomist(s) on the contractual specifications

In the perspective of an efficient collective handling of contractual aspects of HFE under the constraints of costs and delays:

- "Ergonomics material standard" would better explicitly stated as a shared resource by default and stated as requirements to be demonstrated only for mandatory reasons;
- The technical designers would rather be directly in charge of the "material-oriented" contractual requirements related to HFE. This includes the specific ergonomics requirements included in the rules of the classification society but also any specific technical requirements explicitly requested by the client. The technical designers should have the opportunity to get the support from an ergonomist, in case of doubt on the interpretation of some requirements or in case of a lack of experience in the use of ergonomic "material standards"; and
- The ergonomist(s) would rather be responsible for the management and the development of the users-centered approach. This approach necessarily involves the relevant technical designers. The scope would be defined according to a stakes analysis about the innovative aspects of the project and also considering the possible impacts of the industrial organization. Some flexibility should be preserved so that the intervention can be orientated on the most valuable topics that will raise with no major contractual re-negotiation.

If agreed this allocation of responsibilities, then the contractual documentation would better be structured in accordance with it in order to avoid unnecessary interfaces. This would favor the collective efficiency.

Q&A of HFE aspects

If agreed the above considerations and the above suggestions about how to contract material requirements and methodological requirements in a collective strategy, then:

- Demonstrating any generic ergonomic "Material standard" sentence by sentence at whole ship level should be avoided. It is costly regarding the expectable benefits. The wish for a total control through a process of requirements allocation harsh realities of the real world. Such systematic process including compliance matrix over each paragraph or sentence of the standards seems unreasonable at whole ship level. And it does not preserve from a risk of poor design as a compilation of "ergonomic elements" does not necessarily makes an "ergonomic situation". It seems better to invest in the holistic approach focused on the users 'activities;

- The demonstration of (rather) specific HFE contractual material requirement from the rules of classification societies refers to the existing practices; and
- The demonstration of the UCD process should not be neglected. A track from the CYCLADES project is a certification process [Ref. 5], a bit like current practice regarding the design of cockpits in aviation industry. Another track may consist in delivering a kind of label compliant entities, so that the clients can evaluate before signing any contract how far a ship supplier is collectively skilled in UCD. But at this time it does not exist. So, unless successful shared experience between the client and the supplier, this subject should be discussed early enough in the negotiation phase.

8. CONCLUSION

The experience of the DCNS HF team in implementing UCD principles within major projects shows that this fruitful approach is possible within a ship contract environment.

However this full-range experience is with one client in the frame of a historical relationship that eased the mutual understanding. On the way ahead there is a need for sharing a baseline of practices at whole maritime community level so that contracting and implementing UCD does not request so much effort between new partners.

The direction suggested by R.V. POMEROY & B.M. SHERWOOD JONES around 15 years ago [Ref. 1] was relevant and it is worth pursuing it. The CYCLADE project [Ref. 4 & 5] will stimulate the move towards users-centered design process. But the move is slow and many efforts will probably be required yet in the next years. The most difficult point is probably about Q&A of HFE aspects as it seems to be the main fear regarding UCD.

The development of UCD is to be promoted but ergonomists have to take care that it will not be forgotten that it requires a specific skill in analyzing the work. If not, this may lead to no more ergonomists within design teams. The presence of ergonomists within design teams and technical R&D studies will remain useful as an active actor of the design. Beyond this role, effective practice gives the resources for a continuous improvement of methodologies and tools.

9. ACKNOWLEDGEMENTS

The authors wish to thank the DCNS Human Factors community for encouraging this publication. They also wish to thank the HF specialists of DCNS clients for their incentive strategy, and all the users met across DCNS projects for their feed-back on past or current designs.

The views expressed here are those of the authors and not necessarily those of DCNS.

10. REFERENCES

1. POMEROY & SHERWOOD JONES, Managing the Human Element in Modern Ship Design and operation, *Human Factors in Ship Design and Operation, RINA International Conference, London, October 2002*

2. J.F. CHOLAT, La normalisation en Ergonomie, Etat des lieux, 39ème *Congrès SELF, Genève, 2004*

3. G. LE BOUAR & C. CHAUVIN, Apport et limites de la norme EN ISO 8468 dans la conception des passerelles de navires, 39ème *Congrès SELF, Genève, 2004*

4. Handling of usability requirements from a regulatory perspective, *http://www.cyclades-project.eu/CyClaDes/results/handling-of-usability-requirements-from-a-regulatory-perspective (accessed 08/23/2016)*

5. B. SHERWOOD JONES & M. LUTZHOFT, HCD assessment chapter for "improving operational design of ships" revision 2.0 dated 07/05/2015, , *deliverable n° 4.2 of CYCLADES project*

6. L'ergonome interne dans la boucle de conception, A. Drouin (2004).

7. A. KATARIA & G. PRAETOSORIUS & J.U. SCHROEDER-HINRICHS, M. BALDAUF, Making the case for Crew-Centered Design (CDD) in merchant shipping, *Proceedings 19th Triennial Congress of the IEA, Melbourne 9 - 14 August 2015*

8. J.V EARTHY & B.M. SHERWOOD JONES, Design the human factor – The move to goal-based rules, *May 2005*

9. H. MYLES, The Human Element in Ship Design, *August 2015*

10. Alert! Issue n°22, A need to address human element issues effectively, January 2010

11. G. PRAETORIUS & al,, Increased awareness for maritime human factors through e-learning in Crew-centered Design, *6th International Conference on Applied Human Factors and Ergonomics (AHFE 2015) and the Affiliated Conference, AHFE 2015*

12. J. EARTHY & B. SHERWOOD JONES, Best practices for addressing human element issues in the shipping industry, *May 2010, available on ALERT! Website (accessed 19 August 2016)*

Websites:
- Alert! (http://www.he-alert.org/)
- CYCLADES (http://www.cyclades-project.eu)
- Researchgate (https://www.researchgate.net)

11. AUTHORS BIOGRAPHIES

Jean-Philippe Flohic holds the current position of Human Factors Expert at DCNS. He is responsible for the management and the implementation of Human Factors & Ergonomics studies in ship design and also in R&D projects in DCNS. His experience in this company includes more than 16 years of ergonomic intervention into national and international surface warships and submarines programmes from bidding phases to commissioning.

Céline STALLA holds the current position of Human Factors Specialist at DCNS. She is responsible for the management and the implementation of Human Factors & Ergonomics studies in ship design and R&D projects in DCNS. She is also responsible for the quality and developement of the skills and the performances of HF team at DCNS. Her experience at DCNS includes 8 years of ergonomic intervention into surface warships and submarines programmes. This follows a 2-year experience as ergonomist into the nuclear industry.

THE SELF-ASSESSMENT OF SAFETY CULTURE IN MARITIME ORGANISATIONS

C M Tomlinson, ABS, UK
B N Craig, Lamar University, USA
K P McSweeney, ABS, USA

SUMMARY

The safety culture of an organization has been defined as '...individual and group values, attitudes, perceptions, competencies, and patterns of behaviour that determine the commitment to, and the style and proficiency of, an organisation's health and safety management'[1]. ABS conducted maritime safety culture research for over a decade before publishing the *ABS Guidance Notes on Safety Culture and Leading Indicators of Safety* in January 2012. These Guidance Notes enable ship owners and operators to self-assess their safety culture by surveying the workforce using the questionnaire provided in the Guidance Notes. Self-assessment benefits companies because involvement throughout the process leads to better understanding of the results, and insight in utilizing them to maximum advantage. This practical approach taken by ABS is embedded in a theoretical stance that safety culture (not just safety climate) can be assessed by survey without recourse to supplementation with other techniques, such as interviews or focus groups. This paper describes the ABS approach to the self-assessment of safety culture; examines the theory that underpins that approach; and presents preliminary findings from the research undertaken by ABS and Lamar University.

1. INTRODUCTION

Human error accounts for 58% of major claims in shipping[2] and an estimated 80% of all maritime accidents[3], but most of these errors can be traced back to situational factors; this was the repeated finding of a number of investigations into well-publicised disasters (in a variety of industries) undertaken in the 1980s and 1990s. The investigators found that in the final analysis, the cause of each disaster could not be attributed to an error made by a single individual, but arose from more systemic organizational or managerial flaws[4; 5].

During the 1990s findings such as these increased the focus from individual behaviour to the working environment, and beyond that to organizational practices and philosophy. As a result, in all hazardous industries it is now accepted good practice to have in place a safety management system, and to institutionalize safe working practices through the development of a positive safety culture. The central premise behind these concepts is that ongoing vigilance is necessary in all parts of the organization if operational disasters are to be averted.

Safety culture is the set of characteristics and attitudes in an organization which establishes that, as an overriding priority, safety issues receive the attention warranted by their significance. The primary tool used for assessing the safety culture of an organization is the safety culture questionnaire given to a representative sample of the workforce, if not the whole workforce.

ABS has been conducting research into safety culture, and leading indicators of safety, for more than a decade. The development of this initiative was described at RINA 2011[6]. The *ABS Guidance Notes on Safety Culture and Leading Indicators of Safety* were published in January 2012[7]. The Guidance Notes are fully comprehensive, providing all of the information for maritime organizations to complete their own surveys. This includes:

- Administering the survey
- The safety culture questionnaire
- Statistical analysis with worked examples
- How to interpret the results
- An action plan

It takes about 15 minutes to answer the 40-question safety culture questionnaire that has focused on eight safety factors. There are five questions related to each factor. These questions are expressed as Likert statements that employees are asked to rate. The eight safety factors are:

- Communication (COM)
- Empowerment (EMP)
- Feedback (FDB)
- Mutual trust (MTR)
- Problem identification (PID)
- Promotion of safety (POS)
- Responsiveness (RSP)
- Safety awareness (SAW)

Each of these safety factors is defined in the Guidance Notes. For example, Promotion of Safety is defined as: "Managers lead in promoting safety as a core value in the organization. Visible, active, and consistent support for safety programs and objectives exists at all levels of management within the organization. Managers are seen to be committed to doing what is right, demonstrating their values through their communications, actions, priorities, and provision of resources. This value cascades down through all levels of the organization." The questionnaire is divided into three sections:

- Ship safety
- Health and safety
- Respondent's own job

The demographics section has questions about age, gender, nationality, department, position onboard ship, length of service in the maritime industry; length of service with current employer; and length of service in present position. The types of statistical analysis undertaken are:

- Shipboard versus shoreside analysis – difference in responses between crew on vessels and shoreside staff (who interface with the crew)
- Demographic analysis – difference in responses on the basis of age, gender, nationality, departments, job position and experience
- Within groups analysis combines two or more categories of demographics such as:
    - Job position within nationality
    - Job position within age group
    - Nationality within experience
    - Age group within experience
    - Gender within experience

At the end of the questionnaire, there are four questions to be answered in the respondent's own words (free-text fields). These are:

- What could this company do to improve occupational health and safety?
- What could this company do to improve ship safety?
- What could this company do to improve shore-to-ship safety?
- What questions were not asked in this survey, but should have been?

These free-text responses are sanitized and collated for the client. These often run to 30-40 pages and prove to be very interesting. Often the free-text responses are significantly more negative than the responses to the 40 questions, even when covering the same topic.

Clients can create and add their own questions, including free-text questions, if desired. Thus far the questionnaire has been translated into Spanish, Portuguese, Arabic, Mandarin and Tagalog.

2. FUNDAMENTALS OF SAFETY CULTURE

This section addresses some theoretical aspects of the work. Firstly, why is it necessary to carry out safety culture surveys, when what is required is an improvement in behaviour? Is it not easier to observe behaviour, identify weaknesses and devise a program to accomplish that? Secondly, is it even possible to assess safety culture via survey? Or as what is captured is only a snap-shot of safety culture, does that constitute something more transient, commonly called safety climate?

2.1 WHY NOT OBSERVE BEHAVIOUR?

The primary concern about a poor safety culture is that it fosters unsafe practices and behaviour, so it is sometimes argued that observation of behaviour should form part of the assessment of a safety culture. For example, a proposal submitted to IMO to update STCW competencies tables to include five key elements of safety culture, identified the elements as: understanding, behaviour, compliance, risk management and leadership[8]. However, drawing conclusions from observed behaviour is not straightforward. People naturally want to be seen in their best light, so there is a motive to modify their behaviour when they know they are being watched, which can render observation data unreliable.

Furthermore, behaviour is open to interpretation. Social scientists have discovered that culture is acquired in three stages[9]:

1. Conformance (imitation)
2. Identification with others' expectations
3. Internalization (full acquisition)

The first stage is characterized by imitation of others' behaviour without understanding it; the second stage is reached when the person identifies with the expectations of significant others and chooses to comply with them, often because there is a reward for doing so, or a sanction for failure (often referred to as a carrot and stick approach); and, at the third (and final) stage the norms of the culture become second-nature, having been fully internalized.

As the same behaviour can occur at all three stages (for different reasons) the results from observation studies must be cross-checked with probing questions about the reasons for the behaviour, to ascertain which degree of understanding has been reached. Without those checks, little can be learned from behavioural assessments - except where they are undertaken in a simulator environment which can be manipulated to panic and pressure individuals into revealing their inherent behaviour.

It is not possible to improve safety culture by behaviour modification programmes either because that can only force compliance to stage 2. Progress beyond Stage 2 to stage 3 denotes a change of attitude and understanding that gives rise to an altered behaviour. It may be argued that the change in behaviour will lead to a change in attitude, but this is questionable, and even if true, is a dubious way of winning hearts and minds, which is central to the safety culture ethos. The alternative, the survey of safety values, attitudes and perceptions which underpin desired behaviour, is less onerous and a more robust indicator of safety culture maturity as safety attitudes develop with it.

2.2 CULTURE VERSUS CLIMATE

Cultural anthropologists define culture as a system of shared meanings and shared understandings. According to Schein (1992)[10] it comprises three levels:

1. Most superficially, the *artefact level* - containing the visual products of the group
2. Secondly, the *espoused values* - seen in strategies, goals and philosophies
3. At its core, the *basic underlying assumptions* - unconscious, taken-for-granted understandings, assumptions and values

The two terms safety culture and safety climate are often used interchangeably[11] as there is no consensus about the differences between the two concepts. For example, the HSE's website advertises its Safety Climate Tool as "Our Safety Climate Tool gives you a unique insight into your organization's safety culture, and evidence-based suggestions for improving it."[12].

However, it is the norm to argue that only the unconscious assumptions at level 3 constitute safety culture, and the attitudes and views expressed by employees as well as the tools and techniques an organization uses for managing safety (level 1 and 2) are something more transient called safety climate[13].

It is self-evidently true that it is not possible to assess core culture (level 3) by survey in a primitive society; no anthropologist would attempt it. The modus operandi of anthropologists is a longitudinal study, lived within the community being studied, often a primitive tribe with a language not spoken by the anthropologist. The research is written up as ethnography (the anthropologists' principal tool) and analysed using Grounded Theory and associated tools and techniques in qualitative data analysis.

But what needs to be borne in mind is that studying safety culture in the maritime industry is not akin to the research done by anthropologists. This is because the researchers and staff in the maritime industry share a common culture; what is being researched is not a culture but a *subculture*. This is the work, not of anthropologists, but of sociologists (whose scholarship is in the study of subcultures within western industrialized societies), and the primary tool of the sociologist is the questionnaire, used to elicit responses about perceptions, values, attitudes etc. in a multitude of situations.

3. THE SURVEY

The *ABS Guidance Notes on Safety Culture and Leading Indicators of Safety* provides information on how to prepare for and execute the survey. Some key points are:

- When will the survey run? For how long? Starting and finishing dates? Bear in mind that one month should be allowed for preparation, and there should be sufficient time for staff rotation, if appropriate who will be taking the survey? Crew on ships? Crew on leave? Shoreside staff?
- Where will they be taking the survey?
  - Offices
  - Crew on board – do they have access to the internet with adequate bandwidth?
  - Crew on leave – do they have internet access at home? Or do they go into the office?
- Who will be leading the project?
- Preparing staff for the survey.
- Preparing the questionnaire.

4. THE ANALYSIS TOOL

Even with the step-by-step guidance on statistical techniques and worked examples in the Guidance Notes, ABS recognized that many clients would welcome computerized assistance with the task. To this end, ABS and Lamar University have produced an accompanying analysis tool. The analysis tool:

- Includes the safety culture questionnaire as a web-based instrument and in Excel format
- Allows for the inputting or importing of data, and data exports
- Runs the statistics for the safety culture survey
- Helps generates reports
- Stores the data for trending results over time
- Enables benchmarking

The results are presented to the client by a member of the ABS team, usually the ABS project manager. The client has approximately one week to review the reports ahead of the meeting. The reports are generated by the analysis tool, and hand-finished. For the safety culture survey, there are two deliverables:

- The Main Report (~35 pages), and
- The Detailed Findings (~140 pages).

5. THE DELIVERABLES

5.1 THE MAIN REPORT

This report has an executive summary, and contains details of what was done, and an overview of the survey results in tabular form:

1. The five greatest opportunities for improvement based on the shipboard results
2. Five areas of excellence based on the shipboard results
3. The five greatest opportunities for improvement based on the shoreside results
4. Five areas of excellence based on the shoreside results
5. The largest difference between shipboard and shoreside results

The results of all questions are presented within each safety factor and color-coded red, yellow or green (a traffic light system). ABS does not give specific advice on next steps but the report does identify the actions to take to benefit from the survey, for example:

1. Study the findings, and note the safety factors that need to be addressed.
2. Look at the appropriate safety factors' desired activities, attitudes, and behaviors as well as possible activities for improvement to help to decide what to do.
3. Consider if the findings (i.e., significant questions in the safety culture survey) could relate to a different safety factor, as there is some overlap.
4. Prioritize the opportunities for improvement. Initially identify three to five key areas to focus on and develop an action plan.
5. Consider how those key areas align with other initiatives/needs. Focus on strategies that may address more than one area or need.
6. Communicate the results to the workforce. Feedback should include strengths as well as areas of weakness. This can be done in a variety of ways (e.g., written reports, team briefings).This feedback is important for buy-in to safety culture improvement.
7. Engage key shipboard personnel (front-line personnel) in the planning and the trialing of process changes. Action plan development and implementation are typically more successful if these personnel are included.
8. Track changes for continual improvement efforts.

5.2 THE DETAILED FINDINGS

This report expands on the findings given in the report. It also explains the statistics used, and there are ~100 pages of tables with findings from the 'within groups analysis'. This level of detail is usually not required by senior management but is of high interest to the client's project manager and safety manager.

5.3 FREE-TEXT RESPONSES

The client also receives the responses to the 4 free-text questions after they have been sanitized of personal/identifying remarks, and collated. For example, some responses have named these items as requiring attention:

- More and improved training
- PPE availability and quality
- Improved medical treatment on ships and medical exams for crews
- Need to improve hand-over processes
- Work/rest schedule
- Improvement of onboard living conditions

5.4 TRAFFIC LIGHT SYSTEM SET OF SLIDES

Clients are also presented with a Traffic Light System set of slides. Average survey responses (from 1=disagree to 5=agree) are colour-coded and presented in a Traffic Light System format, as shown in Figure 1.

4.5–5.0
4.0-<4.5
1.0-4.0

Figure 1: Scoring of Responses

6. PRELIMARY RESULTS

The analysis tool generates a Report of approximately 35 pages, and Detailed Findings of approximately 140 pages. The Report presents the results in terms of:

1. Opportunities for improvement – shipboard
2. Opportunities for improvement – shoreside
3. Areas of excellence – shipboard
4. Areas of excellence – shoreside
5. Greatest difference in responses between shipboard and shoreside

The results are given in terms of the safety factor as shown in Figure 2, which shows areas of excellence as identified by shipboard responses. There is a similar table for shoreside responses.

Likert Statement	Average Response	Safety Factor
I fully understand my responsibilities for health and safety	**4.96**	MTR
Crew members are actively encouraged to improve safety	**4.94**	EMP
I always ask questions if I don't understand the instructions given to me, or I am unsure of the relevant safety precautions	**4.94**	COM
Safety is the top priority for crew onboard this ship	**4.94**	SAW
The crew has access to all necessary personal protective equipment (PPE)	**4.92**	RSP

Figure 2: Areas of Excellence – Shipboard

Figure 3 shows the results for areas that need improvement as identified by shipboard responses; and a similar table is produced for shoreside responses.

Likert Statement	Average Response	Safety Factor
Shoreside managers never put schedule or costs above safety	3.85	POS
Language differences in multi-cultural crews are not a threat to safety	**4.11**	COM
There are no differences in the performance of crew members from different cultures	**4.24**	MTR
An effective anonymous reporting system exists in this company	**4.49**	COM
This company has excellent maintenance standards	**4.49**	POS

Figure 3: Opportunities for Improvement – Shipboard

Collating the results of the first five clients can give indication of the initial findings. These cover a total of 251 shoreside staff, and 4,205 seafarers on 207 vessels, and are shown in Figure 4.

Likert Statement	Average Response	Safety Factor
There are no differences in the performance of crew members from different cultures	3.40	MTR
Language differences in multi-cultural crews are not a threat to safety	3.43	COM
Mistakes are corrected without punishment and treated as a learning opportunity	3.82	MTR
An effective anonymous reporting system exists in this company	**4.00**	COM
Watch hand-overs are through and not hurried	**4.12**	SAW

Figure 4: Opportunities for Improvement – Combined Results of Shoreside

It has been striking that the answers to the free-text questions at the end of the questionnaire (see section 5.3) are invariably more negative than the Likert Statement responses. Perhaps crew and staff find it easier to give vent to their feelings when they have the opportunity to express them in their own words. There are a number of themes which are shared by both shipboard crew and shoreside staff. These are:

- Training/instruction (for new crew, before changing jobs, hands-on drills, and in general refresher training
- Safety training while on shore leave
- Increased safety meetings and publications
- Increased safety communication
- Increased communication in general (onboard and ship to shore)
- Implementing a just culture
- Improvement of safety culture
- 'Close the loop' on ship-shore safety issues
- Keep crews up to date on company and regulatory issues

7. UTILIZING THE RESULTS

The *ABS Guidance Notes on Safety Culture and Leading Indicators of Safety* includes a list of desired activities, attitudes and behaviours, together with a list of possible activities for improvement for each of the eight safety factors. For example, the desired activities, attitudes and behaviours for Empowerment are given as:

1. The authority and responsibility every employee to terminate a task or activity if there is legitimate safety concerns.

2. All crew feel able to voice concerns and to make suggestions to improve safety.

3. Every employee tries to improve safety.

The following activities are suggested for possible improvement:

- Confirm that all employees know how to voice concerns, whether proactively as an opportunity for improvement or reactively as notice of deficiency
- Solicit workforce opinions on effective communication means and frequency
- Verify that employees have the resources necessary to satisfy safety responsibilities
- Clearly define accountability for safety systems
- Celebrate employee safety decision/action successes
- Institute an off-the-job safety program
- Promote highlighting of employee safety concerns through formal and non-traditional means
- See that employee safety concerns and suggestions are resolved in a credible, timely fashion
- Establish safety committees that include a vertical slice of the organization
- Establish clear, documented accountabilities for safety
- Create an anonymous safety issue reporting system.

This list can be consulted when following the action plan below. The value of the leading indicators process cannot be realized until the results are incorporated into the organization's continual improvement program.

7.1 ACTION PLAN

The following action plan should be implemented to benefit from a leading indicators assessment:

- Review the findings to identify the safety factors (categories of statements or metrics) that need to be addressed
- Look at the appropriate safety factor which contains desired activities, attitudes, and behaviours as well as possible activities for improvement
- Consider if the findings could relate to a different safety factor, as there is some overlap; in that case, consider the desired activities, attitudes, and behaviours and possible activities for improvement for that safety factor as well
- Communicate the results to the workforce - feedback should include strengths as well as areas of weakness; this can be done in a variety of ways, e.g., written reports, or team briefings
- Prioritize the opportunities for improvement; initially identify three to five key areas to focus on and develop an action plan
- Consider how those key areas align with other initiatives/needs
- Focus on strategies that can address more than one area or need
- Engage key shipboard personnel (front-line personnel) in the planning and the trialling of process changes as action plan development and implementation are typically more successful if these personnel are able to be included
- Track changes for continual improvement efforts

8. LESSONS LEARNED TO DATE

Many lessons have been learned from the trials regarding how ABS and LU can improve the process, and these are being implemented, including:

1. The company should appoint a project manager who is enthusiastic about the project – this increases the rate of successful outcomes.

2. The project manager should attend the initial meeting with ABS to discuss the issues involved in participation. In some cases this meeting has been attended by the CEO (because the project manager has not yet been appointed) and the issues relayed second-hand to the project manager. This invariably leads to misunderstandings.

3. The project manager should be allocated sufficient time and resources to do the job. Most companies say that they underestimated the effort involved in self-assessment.

4. The project manager needs to ensure that all data provided is correct. For example, the list of names of departments or job titles should be complete to avoid confusion and rework.

5. Examples of the forms to be completed should be presented to the project manager at the initial meeting, as explaining form-filling by email can be open to misunderstanding.

6. The company should follow the advice given in the Guidance Notes (Chapter 2 Section 2) in preparing for the survey.

7. Clients value the presentations of the results as much as the reports.

8. The survey must be allowed to run for a sufficiently long period of time to capture full rotations if a client wishes to include crew on shore leave in the survey.

9. Shoreside staff that interface with crew should be included in the survey as they often have a different perception from crew.

10. The free-text responses are of far more benefit than was originally anticipated.

11. It is important that the crew and staff know that their responses will be treated confidentially; that individual questionnaires cannot be seen by company shipboard or shoreside personnel

9. CONCLUSIONS

This approach affords clients some impressive benefits:

1. *Ownership (and understanding) of the approach and methods used.* Clients say that they understand the results much more (than previous surveys done by consultants) because they have been involved throughout the process. Consequently, they feel empowered to utilize the results, and even run a further survey independently at some future time. This is testament to the benefit of being an active participant in the process.

2. *Comprehensive instructions and materials.* The ABS Guidance Notes are very comprehensive containing all the material required i.e. questionnaires, datasheets, checklists, metrics, statistical analysis and worked examples.

3. *Analysis tool that semi-automates the process.* As well as containing the questionnaires, the tool carries out the statistical analysis and generates the reports.

4. *Innovative statistical approach to identifying leading indicators from metrics or safety culture surveys.* This is the first statistical approach to leading indicators of safety.

5. *Improve safety performance.* The central premise of the ABS model is that improving safety culture, and/or running a leading indicators program should lead to improvements in safety performance.

6. *The approach allows trending and benchmarking.* Companies will be able to carry out surveys every few years to assess progress, and to compare their results with other (anonymous) companies in the database.

10. REFERENCES

1 ACSNI Study Group on Human Factors, report: Human Reliability Assessment – a Critical Overview. Advisory Committee on the Safety of Nuclear Installations, 1991.

2 UK P&I CLUB Analysis of major claims, ten year trends in maritime risk, UK P&I CLUB, 1999.

3 Lord Donaldson Safer ships, cleaner seas. Report for the UK Government, 1994.

4 Pidgeon N. and Turner B.A. Human error and socio-technical system failure. In: A.S. Nowak (ed.) Modeling human error in structural design and construction. NY: American Society of Civil Engineers, 1986.

5 Leveson N.G. Safeware: System safety and computers. Addison-Wesley, USA, 1995.

6 Tomlinson C.M., Craig B.N., and Meehan, M.J. Enhancing Safety Performance with a Leading Indicators Program. Presented at the RINA Conference, 16-17 November, London, UK, 2011.

7 ABS Guidance Notes on Safety Culture and Leading Indicators of Safety; updated February 2014. Available from the ABS website: www.eagle.org 2012.

8 MEPC\56\17\4 Joint submission to IMO by ICS, IAPH, CEFIC, OCIMF, INTERTANKO and IPTA, 2007.

9 Asch S.E. Social Psychology. Englewood Cliffs, NJ: Prentice Hall, 1952.

10 Schein E.H. Organizational culture and leadership. 2nd edition. Jossey-Bass, San Francisco, 1992.

11 Flin R., Mearns K, O'Connor P. and R. Bryden, Measuring Safety Climate: Identifying the Common Features. Safety Science Vol.34 pp177-192, 2000.

12 HSE's Safety Climate Tool. See the HSE website: www.hsl.gov.uk/products/safety-climate-tool

13 Glendon A.I., Clarke S.G. and E.F. McKenna Human Safety and Risk Management. 2nd edition. Taylor and Francis, 2006.

11. AUTHOR BIOGRAPHIES

Christine Tomlinson, PhD, is a Staff Consultant in the Safety and Human Factors group of ABS. Dr Tomlinson has thirty years' experience of researching how human and organizational issues impact safety-critical systems. Since 1995 she has served on numerous safety-related committees including INTERTANKO's Human Element in Shipping Committee; the MCA's Human Element Advisory Group; and as Human Element Advisor to the Royal Institution of Naval Architecture's IMO Committee. Since 2008, she has been the project manager, and technical lead, for a series of ABS projects looking at safety culture in the maritime and offshore industries.

Brian Craig, PhD, PE, CPE, is Professor/Chair of the Department of Industrial Engineering, and the Director of the Mariner Safety Research Initiative, and the Director of academics and research for the Center for the Advances in Port Management. Dr. Craig's expertise lies in the fields of human factors, ergonomics and safety (HFES) in the maritime and offshore industry, among others. In collaboration with national researchers, practitioners and industry partners, he has (co)developed models, publications, training materials and procedures, etc. that enhance maritime and offshore design, usability and safety performance. Dr. Craig holds a PhD in Interdisciplinary Engineering (specializing in human factors, ergonomics and safety), a MS in Industrial Engineering, a BS in Industrial Engineering, and a BS in Bioengineering all from Texas A&M University, College Station. He has (co)authored/presented over 40 scientific journal, refereed conference proceedings, and worldwide industry guidance documents on maritime and offshore safety performance, practice, and design and is frequent speaker at shipping and offshore industry meetings. He has been PI on over 25 industry and governmentally supported HFES research projects totalling over $1.5M, mostly in the maritime industry.

Kevin McSweeney, PhD, CPE, is the Director, Advanced Technology & Research for Human Factors, Safety, and Risk at ABS. Key responsibilities include: providing the maritime industry with guidance on mariner safety, safety culture, HFE program development and implementation, crew habitability and accommodations design, occupational safety, safety program implementation, the application of ergonomics to vessel and installation designs, identification of leading indicators of safety, and improvement of safety management systems. Kevin is also a Board Certified Professional Ergonomist (CPE).

EARLY HUMAN FACTORS IN COMMAND AND CONTROL SPACE DESIGN

C. Launchbury, BMT Defence Services Ltd., UK.
J. Campbell, BMT Defence Services Ltd., UK.

SUMMARY

The aim of this paper is to highlight the value achieved by considering Human Factors relating to the design of command and control spaces early in the Concept phase. This stage of design involves numerous iterations influenced by competing engineering pressures and continuous compromises and trade-offs; compartments frequently change size, shape and location with corresponding changes in position of doors, hatches and stairs. It is at this early stage where influence from Human Factors specialists is required and where design changes for the benefit of the end user can be most easily made. The aim of this work was to enable BMT Defence Services Ltd Human Factors team to effectively represent the end user in design trade-off decisions during the Concept phase of BMT's Venator®-110 light frigate. This was achieved by applying an efficient user-centred approach informed by established guidance but applied earlier than the current guidance suggests. The result is a more effective vessel taking account of end user requirements more cost effectively than would otherwise have been the case.

NOMENCLATURE

BMT DSL	BMT Defence Services Ltd.
Dstl	Defence Science & Technology Laboratory
EHFA	Early Human Factors Analysis
GA	General Arrangement
HF	Human Factors
MAP	Maritime Acquisition Publication
MOD	Ministry of Defence
SME	Subject Matter Expert
UK	United Kingdom

1. INTRODUCTION

1.1 BACKGROUND

The work described in this paper contributed to the Concept phase for the development of a BMT Defence Services Ltd (BMT DSL) light frigate, known as Venator®-110. The vessel is designed to fulfil a comprehensive range of roles from maritime security and boarding operations, to providing credible offensive and defensive warfighting capabilities. BMT's Human Factors specialists worked closely within a multi-disciplinary project team, specifically focusing on the design and layout of the Operations Room and its implications on the General Arrangement (GA).

The benefits of considering human factors early in the design process are well understood. Key Human Factors standards within the UK Ministry of Defence (MOD), for example, identify the importance of establishing and managing Human Factors activities as early as possible once a project team has been established [1]. Many studies have shown that the consideration of the user at the earliest opportunity can improve operational performance, reduce the likelihood of expensive downstream redesign and mitigate key risks [2].

During the basic design of the GA, especially early in the Concept phase of a new vessel, there is a need to consider continuous compromises and trade-offs between competing engineering pressures. This can lead to many design iterations, where individual compartments may change size, shape or location with corresponding changes in the position of doors, hatches and stairs. It is at this early stage where Human Factors can make the greatest contribution and where design changes for the benefit of the end user can be most easily made. For example, the optimal arrangement of consoles within a compartment impacts the ideal location of doors, passageways and stairs.

Established Human Factors guidance for the design of command spaces [3] and naval Human Factors Integration [4] [5] are valuable sources of good practice. However, there is currently no requirement to consider details to the level of individual console layout and traffic flows *within* compartments at such an early stage. Typically, the detailed design of command spaces is a lengthy, expensive process begun later in the Assessment phase, once the GA is mature and space allocation has been determined [3]. At this point it is often prohibitively expensive to make any significant changes to the GA.

This paper presents an overview of work carried out to develop a human centred design. We demonstrate the benefit of an abridged and cost effective command space design process in order to influence the GA as early as possible.

1.2 STUDY AIMS AND SCOPE

The aim of the project was to review, agree and justify the space allocated to the Operations Room and identify any associated user requirements impacting the GA.

At the time BMT Venator®-110 was early in the Concept phase, the GA was not firmly established, the initial role to equipment mapping had not been completed and there was a limited number of defined user requirements. There was also limited funding for a full Human Factors analysis, as recommended by Maritime Acquisition

Publication (MAP) 02-001 [3]. Therefore an abridged, more cost effective version of the full process was developed, which influenced the GA and ensured the Operations Room space was designed with future user considerations at the core.

The study did not aim to identify the manpower requirements to underpin the ship. Expected Operation Room manning numbers were provided by Subject Matter Experts (SMEs) and based on informed opinion to be validated through ongoing manning studies. In addition, the specific equipment to be housed in the Command Room had not been specified.

2. METHODOLOGY

A working group comprising Human Factors specialists, Naval Architects, ex-Royal Navy ship operators and engineers, and 3D visualisation modellers was set up to support the project. Each role brought a different perspective and priority as to what was required from the Operations Room and how it would be used. This multi-disciplinary team allowed the exploration and balancing of different requirements and constraints.

In order to guide the project methodology, the study team initially conducted a literature review of current guidance on Operations Room design, involving maritime standards, Human Factors Integration (HFI) standards and ISO design standards.

Command and control space design is the focus of a dedicated MOD MAP - 02-001 'Design of Warship Command and Control Spaces (2013)' [3]. It provides a thorough process model on how to design and develop command and control spaces and the associated assessments required. However, the guidance is aimed at the Assessment phase in the capability acquisition lifecycle where the initial GA is provided as a fixed input. It therefore does not provide guidance for determining the shape and size of the command space itself.

MAP 01-011 'HFI Technical Guide' provides high-level guidance for Human Factors activities during the Concept phase. However, this concerns the arrangement of compartments in relation to each other, including for example, traffic flow analysis between compartments, rather than any detailed requirements to define command spaces. Again, consideration of the layout within the Operations Room is left until the Assessment phase once other design pressures have already defined the available space.

A variety of other standards were also reviewed to identify the key methods to apply to the early design of the Operations Room ([6],[7],[8]). ISO 9241-210:2010 'Ergonomics of human-system interaction' discusses the importance of having user involved early in the design process. The main emphasis is on the application of an iterative design and assessment approach to ensure the design is informed and refined through user evaluation and feedback.

Given the fluidity of the GA and the low level of available detail in customer requirements and equipment fit, a streamlined process was developed based on MAP 002-001. This started with an Early Comparability Analysis to learn from previous designs, and an Early Human Factors Analysis (EHFA) to identify key human related considerations. To reduce costs, a link analysis was conducted using subjective opinion of SMEs rather than detailed observation. Next, the multi-disciplinary team used an iterative design process to generate a range of layout options within constraints imposed by the available space. Each candidate design was assessed using a Multi-Criteria Decision Analysis. The highest scoring layout was then used to provide feedback and identify improvements on the GA.

Throughout the design, the Human Factors team consulted closely with SMEs, who had operational end-user experience working within command spaces on a range of ship types.

3. DESIGN PROCESS

3.1 EARLY COMPARIBILITY ANALYSIS

An Early Comparability Analysis of similar Operations Rooms was conducted with SMEs to identify both strengths and weaknesses of comparable designs and to establish any best practice that could be assimilated into the new design. The focus was on the Royal Navy Type-23 frigate, the Type-45 Destroyer and other European designs with similar roles. This analysis was fed into the EHFA and helped support the requirements definition.

3.2 EARLY HUMAN FACTORS ANALYSIS

An EHFA is a key element in the MOD Human Factors Integration process. It provides a systematic risk-based framework for identifying, tracking and mitigating human related considerations ([9],[10]).

An EHFA workshop was conducted with stakeholders, experienced end-users and representatives with a variety of relevant expertise. The seven domains of HFI (Manpower, Personnel, Training, Human Factors Engineering, System Safety, Health Hazard Assessment, Social and Organisational) were used as a framework to identify key human related considerations. All considerations were recorded in a RAIDO register (Risks, Assumptions, Issues, Dependencies and Opportunities). Each consideration was assessed for likelihood and severity of impact and used to inform the subsequent design process.

3.3 SUBJECTIVE LINK ANALYSIS

Link Analysis is used to better understand the communication between different operators and to determine the frequency of use and importance of each line of communication [3]. In Naval ships' Operations Rooms, verbal communication is supported through the use of intercom headsets. However, physical arrangement of role-specific consoles is an essential consideration affecting the quality of communication using other modalities such as visual communication where face to face or screen/equipment sharing is required.

Ideally, this analysis would be conducted through observation of a range of representative end users over a number of different scenarios. However, it is not economically viable, or helpful, to conduct this level of analysis given the uncertainty regarding the assumed customer requirements. Therefore, it was decided that a more appropriate level of analysis could be achieved (in terms of speed, cost and accuracy) based on the subjective opinion of experienced SMEs.

Rather than consider the detail (modality, frequency or importance) of communication flow between roles, six SMEs were simply asked to rate on a scale of 1-10 how important it was for one role to be in close physical proximity to another role. Given the project preference for low risk and high maturity, the assumption was made that the equipment fit used within the operation room would be similar to current generation, with which the SMEs had experience. At the end of the interviews a workshop was held with all SMEs to consolidate and agree the matrix of scores.

3.4 ITERATIVE DESIGN PROCESS

The Human Factors team were presented with a complete (yet flexible) GA with a reserved space for the Operations Room, alongside neighbouring compartments, passageways, stairs etc. The initial shape and dimensions of the space was influenced by engineering considerations (such as structural bulkheads, service trunks and airflows) and the experience of Naval Architects from previous projects.

Considerations identified by SMEs during the EHFA regarding the working practices and ideal working environments quickly led to several adjustments to the GA. This included the shape of the space, use of compartments and the locations of doors, stairs and passageways.

Next, user requirements and the link analysis were used to generate a wide range of conceptually different layouts in order to avoid converging on an Operations Room layout too early. The aim was to rapidly consider a broad range of alternatives rather than to focus on the detail of a narrow range (for an example, see figure 1 below).

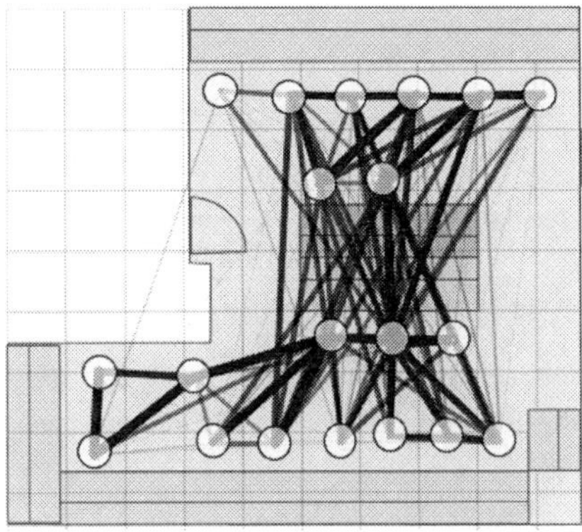

Figure 1: One of seven layouts guided by link analysis

An SME workshop was held to further develop the detail of candidate layouts. Each layout was presented as a 2-dimensional plan using a print out of the GA and movable paper consoles (see figure 2). A number of criteria were considered, such as compliance with link analysis, sightlines, traffic flows and inclusion of major ancillary planning items (as described in MAP 02-001). Weaknesses in each design were identified and improvements made. This was conducted with multiple SMEs to create informed discussion and reduce bias; SMEs were asked to challenge each other to establish a justification for the design.

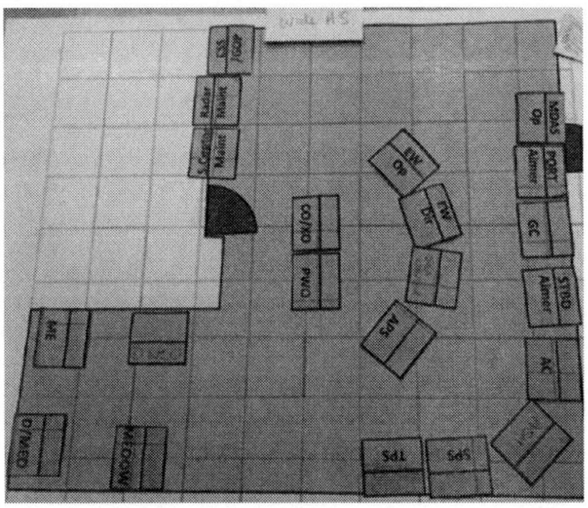

Figure 2: A proposed layout

3.5 MULTI-CRITERIA DECISION ANALYSIS

In order to objectively compare different competing layouts, a simple Weighted Sum Model scoring matrix was used (Multi-Criteria Decision Analysis) [11]. The criteria used within the iterative design process can themselves be subject to detailed analysis. However, to arrive at a sufficiently accurate answer to allow comparison, each criterion was subjectively scored during a workshop of SMEs and Human Factors specialists. Weightings to provide relative importance between criteria were also based on SME opinion. The highest scoring layout was selected as the most suitable candidate for further development.

3.6 3D VISUALIATION

The selected layout was used to generate a detailed 3D model and multiple high-realism visualisations (see figure 3). This allowed the Human Factors team, Naval Architects and the SMEs to review the Operations Room from differing vantage points, resulting in minor adjustments to the equipment positioning.

Figure 3: 3D visualisation of final layout

Anthropometric assessments were conducted to evaluate any constraints imposed by the body dimensions. The 5th percentile female to the 95th percentile male sitting eye heights and standing eye heights (UK Royal Navy) [12] were used. This ensured specific roles (e.g. the Commanding Officer) had visibility of key positions around the compartment (see figure 4).

Figure 4: 5th percentile female view from Commanding Officer's console.

The final layout was then used to elicit further feedback in the design of associated areas of the GA, further influencing the use of space and the placement of doors, stairs and routing of passageways.

4. DISCUSSION

Involvement of Human Factors early in the process is well understood and is practiced in many aspects of military procurement (as defined in Joint Service Publication 912 Human Factors Integration in Defence Systems). The value of using Human Factors analysis as a risk reduction activity, employed at the beginning to identify and avoid user related issues rather than brought in later to solve problems at greater cost once they have materialised is well understood (if not always applied).

However relevant guidance provided by MAP 02-01 [3] is intended for later stages of design by which time the GA design has been influenced by engineering pressures and changes to improve usability are prohibitively expensive.

The purpose of this work was to advise on the shape and size of the space allocated to the Operations Room and associated design implications for the GA. Due to the current lack of guidance regarding space allocation during the Concept phase we adapted and condensed existent guidance to generate a cost effective and rapid process. To this end, selected HFI activities were conducted quickly at a high level with reduced detail. The result provided justification and direction for improvements to the GA, and evidence that the allocated space and shape is sufficient for an effective Operations Room design.

It should be noted that the data collection was carried out with a small number of current and ex-Royal Navy SMEs (n=6), therefore there is a natural bias to the experience and requirements of the Royal Navy. Once the requirements of other customer Navies have been collated (regarding the ship's roles, capabilities, complement and ways of working etc.) the full design process as laid down by MAP 02-001 should be revisited in greater detail. This will include a revalidation of all assumptions and an update of the EHFA.

Having a diverse stakeholder group proved to be beneficial to this early design process; including Human Factors, Naval Architects and modelling specialists, as well as SMEs (current and ex-Royal Navy operators). This diversity allowed competing priorities to be discussed, considered and balanced, and ensured that the design was not dominated or constrained by a single perspective.

It should be noted that under (what was) the Maritime Collaborative Enterprise (MarCE) Dstl have sponsored research looking into the future of operation room design, addressing collaborative working and layout options. This work is currently ongoing, however once complete it will be reviewed and any appropriate recommendations will be assimilated into our command space design process.

Further work is required to validate the current Operations Room design which will include a physical mock-up using functional consoles to validate the link analysis and communication flows. Also, the computer generated 3D model is currently being assessed using virtual reality and motion capture apparatus to validate sightlines, traffic flows and access requirements.

5. CONCLUSION

This study highlights the benefit of including Human Factors specialists as part of the ship design team in the very early stages of design. It provides valuable insights into how space could, and should, be used.

This study also highlights a gap in current guidance regarding analytical activities to be conducted in the Concept phase. The tasks of the end user and the use of command and control spaces should be considered while the size and shape of the space and associated GA can still be influenced.

This work has ensured that Operations Room user requirements have been identified and considered, and have influenced the GA before changes become prohibitively expensive and time-consuming. In doing so, the work also provides customers with confidence that the vessel has been designed around the operational role of the end user, and not compromised by the convenience of manufacture.

6. ACKNOWLEDGEMENTS

The study was funded by BMT Defence Services Ltd. Copyright: BMT DSL. The authors wish to express their thanks to all of the subject matter experts who participated in the study.

7. REFERENCES

1. Joint Service Publication 912, Human Factors Integration in Defence Systems, UK MOD, 2015

2. Bruseberg, A., Cost-Benefit Analysis for Human Factors Integration. A practical guide. UK MOD HFI DTC, 2009.

3. Maritime Acquisition Publication 02-001, Design of Warship Command and Control Spaces. Issue 1. UK MOD, 2013.

4. Maritime Acquisition Publication 01-010 Human Factors Integration Management Guide (formerly STGP 10). Issue 4. UK MOD, 2006.

5. Maritime Acquisition Publication 01-011, Human Factors Integration Technical Guide (formerly STGP 11). Issue 4. UK MOD, 2006.

6. ISO 11064-1, Ergonomic Design of Control Centres. Part 1: Principles for the design of control centres, 2001.

7. ISO 9241-210, Ergonomics of human-system interaction. Part 210: Human-centred design for interactive systems, 2010.

8. Workplace health, safety and welfare regulations, Approved Code of Practice and Guidance, UK Health and Safety Executive, 1992.

9. Defence Standard 00-251, Human Factors Integration for Defence Systems, UK MOD, 2016.

10. Harmer, S., Early Human Factors Analysis Methodology Guide. Issue 1. MOD Technical Note 0-DHCSTC_I2_H_T2_080_002. In press. 2016.

11. Köksalan, M., Wallenius, J., and Zionts, S. Multiple Criteria Decision Making: From Early History to the 21st Century. Singapore: World Scientific, 2011.

12. Human Factors Integration, Technical Guide for Anthropometry: People Size. DA4T&QA, UK MOD, 2015.

8. AUTHORS BIOGRAPHY

Claire Launchbury BSc. MSc. C.ErgHF is a Principal Consultant within the Human Factors team at BMT Defence Services Ltd. Claire has over twelve years' experience, working on military and civil projects within multidisciplinary teams across a wide range of applications. Previously Claire was involved in the MOD Dstl Maritime Collaborative Enterprise (MarCE). More recently, she has worked on several MOD Dstl sponsored studies looking at Royal Navy Engineer expectations.

James Campbell BSc. is a Senior Consultant within the Human Factors team at BMT Defence Services Ltd. He is responsible for delivering Human Factors research and advice to external customers and internal project teams. James has over four years' experience as a Human Factors specialist in the defence sector, contributing to Royal Navy Type 26 Global Combat Ship Human Factors Engineering and several MOD Dstl sponsored studies. Previously James was an officer in the UK Royal Air Force with liaison experience with the Royal Marines, Royal Navy and British Army.

Human Factors, 28-29 September 2016, London, UK

COGNITIVE MODELLING OF CREW RESPONSE DURING NAVAL DAMAGE CONTROL

A Woolley and **J Liu**, Defence Science and Technology Group, Australia

SUMMARY

Understanding crew response to damage events is critical in the analysis of naval platform survivability. Naval Damage Control (DC) decision-making and behaviour is primarily governed by Standard Operating Procedures (SOPs), which provide a simple, logical construct for analysis. Using a technique known as Hierarchical Task Analysis (HTA), SOPs used by the Royal Australian Navy (RAN) were analysed to identify the goals and tasks, and associated trigger conditions, that need to be achieved during DC operations. Information represented in the HTA is being implemented as Behaviour Trees to enable rapid prototyping of intelligent artificial agents (representing naval platform crew) imbued with the knowledge inherent in RAN DC SOPs. The intelligent agents will be used to model containment of, and recovery from, damage events. This will facilitate assessment of changes to crew composition and changes to DC SOPs in relation to the survivability of RAN platforms.

NOMENCLATURE

BA	Breathing Apparatus
C&C	Command & Control
CDM	Critical Decision Method
DC	Damage Control
DCC	Damage Control Central
FAA	First Aid Attack
HTA	Hierarchical Task Analysis
IRM	Integrated Recoverability Model
ISSAC	Integrated Ship Survivability Assessment Capability
M&S	Modelling and Simulation
mEX	maritimeEXODUS
NPC	Non-Player Character
RAN	Royal Australian Navy
SMF	SMARTFIRE
SOP	Standard Operating Procedure
WDA	Work Domain Analysis

1. INTRODUCTION

Naval recoverability is the ability to control ongoing damage, and recover system functionality and platform capability after a damage event [1, 2]. Paramount to platform recoverability is the crew and their behaviour. Recoverability operations are time critical, resource intensive activities affecting integrated survivability, platform mission effectiveness and crew safety. Crew decision-making, situation awareness, coordination of response, changes to crew composition (due to incapacitation and fatigue), and changes to Standard Operating Procedures (SOPs) can affect recoverability operations to restore mission critical system functionality. Understanding crew response to damage events is, therefore, essential for the modelling of naval Damage Control (DC) operations and assessing naval platform integrated survivability.

To analyse the Royal Australian Navy (RAN) DC system requires a methodology that accommodates the complexities of RAN DC operations. A suitable holistic methodology would provide an in-depth analysis of crew cognition and interaction during periods of high cognitive and physical stress loadings within RAN DC. For example, consideration must be given to roles and functions of the crew, the distributed cognition during DC operations, and crew inter-team and intra-team interaction. A proposed methodology for the holistic analysis of crew cognition and interaction during RAN DC incidents consists of six steps [2]:

Step 1. Design DC scenarios to be analysed.
Step 2. Define the boundaries of the 'Object World'.
Step 3. Work Domain Analysis (WDA) of the Object World.
Step 4. Perform Event Analysis of Systematic Teamwork component methods:
  a. Hierarchical Task Analysis (HTA);
  b. Critical Decision Method (CDM);
  b. Co-ordination Demand Analysis;
  c. Communication Usage Diagram; and
  d. Social Network Analysis.
Step 5. Perform National Aeronautics and Space Administration Task Load Index analysis.
Step 6. Perform Systematic Human Error Reduction and Prediction Approach analysis.

The primary goal in performing this analysis is to create intelligent, artificial agents for use in DC Modelling and Simulation (M&S). The proposed methodology [2] will enable the development of such an M&S environment. This will provide enhanced understanding of the constraints and factors affecting naval DC situation awareness, decision-making, and team communication and interaction. Subsequently, this will facilitate the evaluation of DC procedures, policies, training programs and the effect of the insertion of new technologies or autonomous recoverability systems.

The first three steps of the proposed methodology for holistic analysis of RAN DC are complete [2, 3], though requiring validation and refinement. The output from Step 4a is being validated and implemented as Behaviour Trees within the UNITY3D simulation engine [4]. These Behaviour Trees will control the actions of the artificial agents within the M&S environment to assess recoverability operations onboard naval platforms. The Behaviour Trees will form the basis of recoverability models that contribute to an Integrated Ship Survivability Assessment Capability (ISSAC) being developed to facilitate assessment of naval platform survivability. ISSAC will then be utilised to assure platform mission effectiveness, ship survivability and crew safety [5]. This paper presents the need for analysis of DC and recoverability operations, and how that analysis will contribute to development of the Behaviour Trees for use in the ISSAC framework.

2. A NEED FOR COGNITIVE ANALYSIS OF DAMAGE CONTROL PROCEDURES

The physics of damage events is reasonably well understood, with models available for analysing (for example) weapon blast damage [6], systems functionality mapped to capability [7], and fire and smoke spread [8]. However, once a damage event has occurred, survivability is a function of the platform's safety systems, crew decision-making, crew response and the subsequent effects on the ongoing DC incident. Consequently, crew response is a critical factor in platform mission effectiveness, platform survivability and crew safety.

M&S tools that simulate aspects of human behaviour in an emergency situation are available. For example, maritimeEXODUS (mEX) [9, 10] is primarily used for analysing the movement of people in the evacuation of passengers and crew from maritime platforms; and the Integrated Recoverability Model (IRM) [11] emulates aspects of crew DC behaviour for naval platform recoverability operations. However, mEX and IRM, in their current states, do not allow for autonomous decision-making or autonomous behaviour based on DC SOPs.

Previous research utilising Event Trees [8] included the duration required for crew to perform recoverability operations, and enabled limited probability analysis in relation to the consequences of those operations. However, generating Event Trees is a time consuming, subjective form of analysis requiring detailed scenario descriptions, and input from experienced RAN DC practitioners. Event Tree usage becomes more difficult when scenario changes are required.

Consequently, recoverability M&S capability is required to enable repeated simulation of DC scenarios, to explore scenario changes, and to analyse the consequences on platform survivability and mission effectiveness. The recoverability capability requires detailed understanding of DC operations so as to imbue intelligent, artificial agents with the ability to perform autonomous DC actions in response to the observed situation.

The RAN DC organisation consists of Command and Control (C&C), DC Central (DCC), Repair Bases and DC teams. Implementing SOPs to control artificial agents representing the crew during simulated DC operations requires models representing the specific levels of the DC organisation. C&C have operational goals to be achieved while maintaining situation awareness of the external environment and making decisions in relation to mission priorities. DCC prioritise DC activities to enable C&C to achieve their goals. The Repair Bases co-ordinate DC activities in their assigned area of the platform; and DC teams are goal driven for the successful completion of the DC activities to which they have been assigned. Governed by the DC SOPs, and as a result of intensive training, the actions of the crew are systematic in nature. For the DC teams, their decisions and actions are governed by the priorities issued by DCC. The DC teams must also provide DC situation status updates, which is assimilated by DCC and forwarded to C&C. C&C will then re-assess the situation and re-issue DC priorities, as required. In the M&S domain, this could be viewed as a 'scheduling' problem, or a 'goal oriented tasking' problem, whereby the models follow an explicit set of rules in response to a damage event. Due to the variability in scenarios, the models will not have prescribed foreknowledge of the decision paths. However, in reality, decisions are dynamically based on the situations encountered by the crew. A simple configuration change to the platform or to the scenario could result in major changes to the dynamic decision path. Understanding how these changes affect platform survivability is the key driver for developing the recoverability M&S capability. Therefore, developing rules for crew DC response is an important feature of the recoverability model since it will facilitate survivability assessment in accordance with SOPs.

Recoverability is an M&S gap in the current ISSAC requirement. To develop suitable recoverability models requires a detailed understanding of RAN DC SOPs. Steps 1, 2 and 3 in the methodology for the holistic analysis of crew cognition and interaction during RAN DC scenarios have elucidated aspects of RAN DC operations and SOPs [2, 3]. The next step, performing HTA, enables identification of the goals and associated tasks to be achieved during DC incidents likely to be encountered by RAN platform crew during recoverability operations.

3. HIERARCHICHAL TASK ANALYSIS

HTA is a comprehensive, descriptive, normative approach to indentify the contextual conditions for the activation of various goals, sub-goals and tasks during the operation of any given system [12]. For RAN DC analysis, HTA was chosen for its applicability and complementary properties to the WDA performed during Step 3 of the holistic analysis of crew cognition and interaction during RAN DC operations [2].

HTA is performed by analysing doctrine and relevant material that describes the system, interviewing experienced users, and from field observation. HTA consists of a goal and task hierarchy, which is constructed by identifying clusters of tasks required to achieve the respective goals. Each sub-goal is decomposed to the desired level of granularity, such that the primitive tasks required to achieve the goal hierarchy are defined (where the most primitive tasks are those that cannot be decomposed further). Associated with each goal is a Plan describing the order of precedence for each sub-goal to be achieved or tasks to be performed. Plans are the control structures that identify conditions to trigger the sub-goals for any super-ordinate goal. Shepherd [13] defined six basic types of Plans: fixed sequences; contingent sequences; choices; optional completion; concurrent operations; and cycles. The differing Plan types take the variety of sub-goal triggers into account, with complex tasks requiring combinations of the different Plan types. The triggers might include, for example: time; environmental conditions; completion of other sub-goals; system state; and/or receipt of information. As well as identifying the sub-goal trigger conditions, it is also important to identify the exit conditions for the Plans. The goals and sub-goals, and the order in which the tasks are performed, may change markedly between different scenarios.

Consider, for example, the simple HTA presented in Figure 1, showing the goals and tasks associated with borrowing a library book. In the example, the primary goal is to 'Borrow a book from the library'. The primary goal is achieved by successfully completing the sub-goals: 'Go to library', 'Find required book', 'Retrieve book from shelf' and 'Take book to counter'. The order of precedence to achieve these goals is defined in Plan 0 by referencing the respectively numbered sub-goals. Drilling down the hierarchy, specific tasks associated with the goals are identified. So, for Goal 2 'Find required book', the person must achieve the sub-goals 2.1, 2.2, 2.3 2.4 and 2.5 in the order prescribed in Plan 2. Further sub-goals and tasks (not shown) would describe how to achieve the super-ordinate goals.

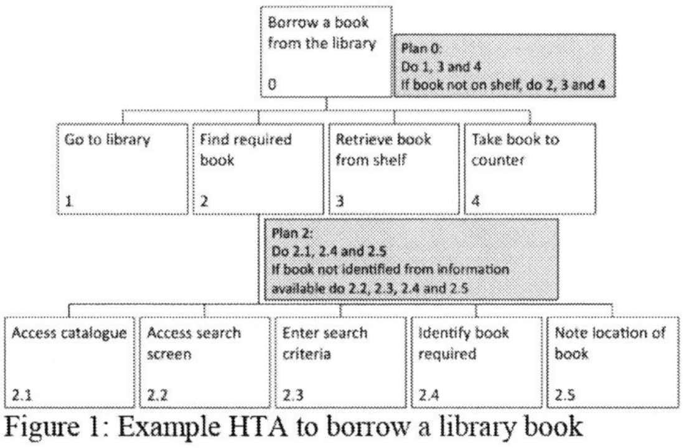

Figure 1: Example HTA to borrow a library book

It is worth noting that sub-goals within the HTA decomposition might afford more than one super-ordinate goal. For example, Goal 2 'Find required book', Goal 3 'Retrieve book from shelf' and Goal 4 'Take book to counter' can also serve the goal of purchasing a book.

4. HTA DECOMPOSITION FOR RAN DAMAGE CONTROL

Presented in Figure 2 is the output from Step 2, the Object World of the domain space, for the holistic analysis of crew cognition and interaction during RAN DC scenarios [2]. The Object World defines the perspective for analysis, constituting everything contained within the confines of the platform and the crew structure. To simplify analysis, the external environment (for example: weather; sea state; outside assistance; and threat space) was not considered at this time.

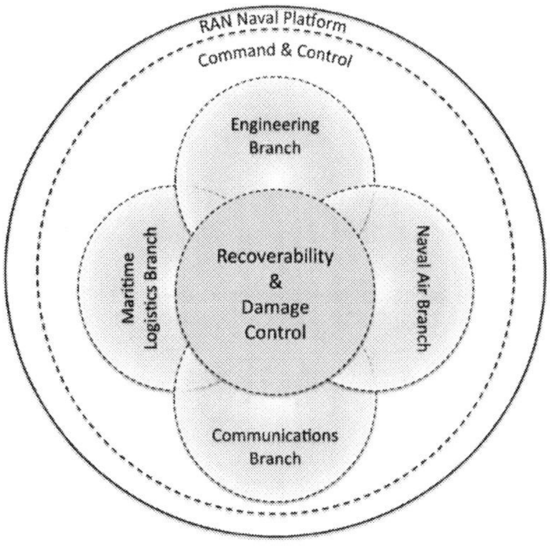

Figure 2: Object World for the HTA domain

HTA was utilised to describe RAN recoverability and DC operations in the context of a hierarchy of goals and tasks within the Object World. Reference material used during the HTA for RAN DC included: RAN Australian Books of Reference detailing RAN combat survivability procedures; DC training aide memoires used during RAN platform work-up periods; field observation; and flow charts depicting RAN DC SOPs [3].

The output from Step 3, the WDA of the Object World for the holistic analysis of crew cognition and interaction during RAN DC scenarios, identified DC objectives and functions [2, 3]. WDA identified the primary goal of RAN DC as 'contain, control and recover from damage', with the priorities of ensuring mission effectiveness, ship survival, and crew safety. To achieve the primary goal, in alignment with the three priorities, there are a number of functions the crew need to perform. These include: fire fighting; flood control; toxic hazard response; and casualty management. Numerous other functions identified within the WDA assist the crew and platform to successfully 'contain, control and recover from damage'; however, they are beyond the crew-focus M&S requirements for ISSAC. Consequently, the HTA was performed on four major categories of recoverability incidents:

- safely control and recover from fire;
- safely control and recover from flooding;
- safely control and recover from toxic hazard incident; and
- manage action casualty during non-mass casualty incident.

Sample HTA decomposition for each of the four DC goals associated with fire fighting, flood control, toxic hazard response and casualty management are presented in Figures 3, 4, 5 and 6, respectively. Note the dashed lines in these figures indicate more details not shown.

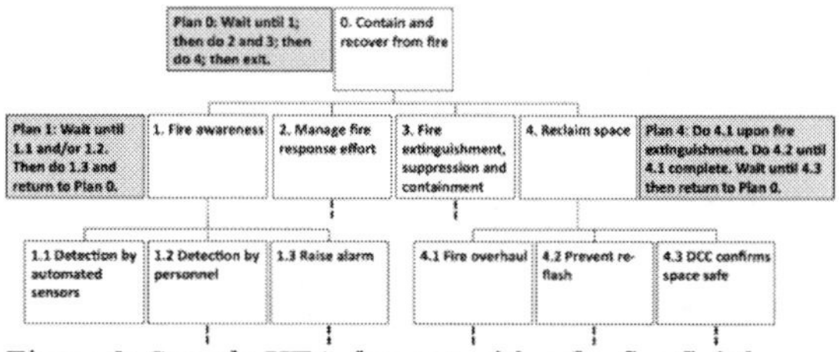

Figure 3: Sample HTA decomposition for fire fighting

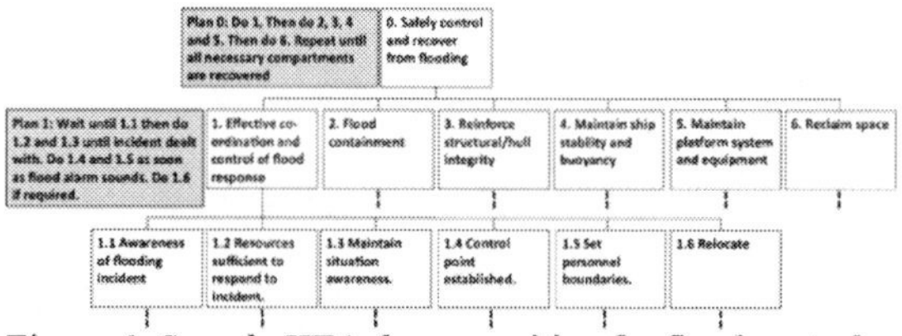

Figure 4: Sample HTA decomposition for flood control

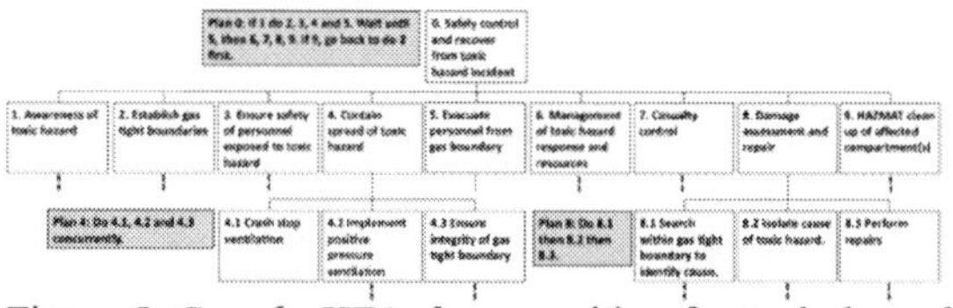

Figure 5: Sample HTA decomposition for toxic hazard response

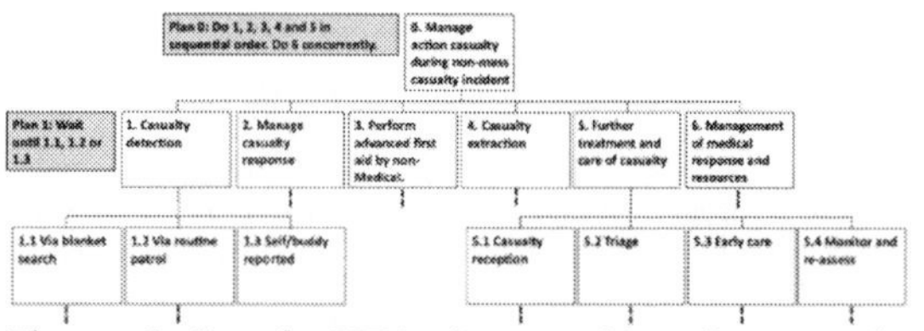

Figure 6: Sample HTA decomposition for managing action casualties during a non-mass casualty incident

The HTA decompositions represent an action plan to be performed when a damage incident occurs. However, during combat operations, there is strong likelihood that various damage incidents will occur at the same time or as a result of an ongoing incident. For example, a detonating missile strike is likely to result in fire and casualties. Therefore, action plans for each of fire fighting, flood control, toxic hazard response and casualty response may be utilised concurrently at numerous times during a DC incident.

Each HTA decomposes into a number of levels making it difficult to present them in their entirety. To illustrate the detail within an RAN DC HTA decomposition and the action plan it represents, consider the HTA decomposition for fire fighting.

4.1 FIRE FIGHTING HTA DECOMPOSITION

The process for fighting a fire onboard a naval platform may be separated into four stages of response, consistent with RAN fire fighting doctrine. The four stages are [8]:

- *Manage* – development of a DC strategy for the incident, maintaining situational awareness, and management of available resources during the incident. The process is initiated by an alarm from the crew or automated systems upon detection of the fire.
- *Extinguish* – upon detecting a fire, the initial aim is to extinguish it. This involves fire teams using fixed and portable fire fighting systems. If the initial attack on the fire is successful, fire extinguishment will be achieved and recovery of the affected compartments and systems is initiated.
- *Suppress* – if the initial attack on the fire is unsuccessful, the aim is to suppress it, including:
  - restricting air supply by closing ventilation systems, doors and hatches; and
  - limiting fuel supply by closing isolation valves and shutting down machinery.
- *Contain* – to limit the growth of the fire within the affected compartment(s) and stop it spreading to surrounding compartment(s). This includes control measures such as: establishment of boundary cooling; and smoke boundaries.

When containment measures are established, fire fighting teams resourced with Breathing Apparatus (BA), fire suits, radio communication and fire hoses are able to attack the fire. Fixed systems may be initiated, where available. Although the fire fighting process may follow the *Manage-Extinguish-Suppress-Contain* stages, each respective stage is not necessarily equal in time; and there may be situations resulting in very rapid succession from one stage to the next. Furthermore, the process may repeat itself after *Contain* when the crew attempts once more to suppress and, then, extinguish the fire. Extinguishment may take some time, during which, resources such as BA may become limited and the attack suspended. The process concludes with reclaiming, or making safe, the space containing the fire.

Flow charts representing fire fighting SOPs for the *Manage-Extinguish-Suppress-Contain* process were developed with the intention of using them as a deterministic rule base for an RAN DC recoverability decision model [3]. However, flow charts do not contain easily identifiable goals and do not explicitly identify action plans. Instead, they represent state transitions in the fire fighting process - documenting the procedural response the crew need to follow when performing RAN DC operations. Conversely, HTA decomposition is purpose oriented - depicting the goals to be achieved and the tasks required to achieve those goals. While the flow charts and the process they represent may be easy to follow, the state representation contained within becomes difficult to implement. In complex simulations, such as managing a combined fire fighting and flooding scenario, traversing the state search space to create action plans becomes time consuming. Conversely, the HTA decomposes an action plan into primitive goals and tasks such that, when implemented in an M&S environment, the creation of actionable plans can be selected from an identifiable, reduced set of tasks [14].

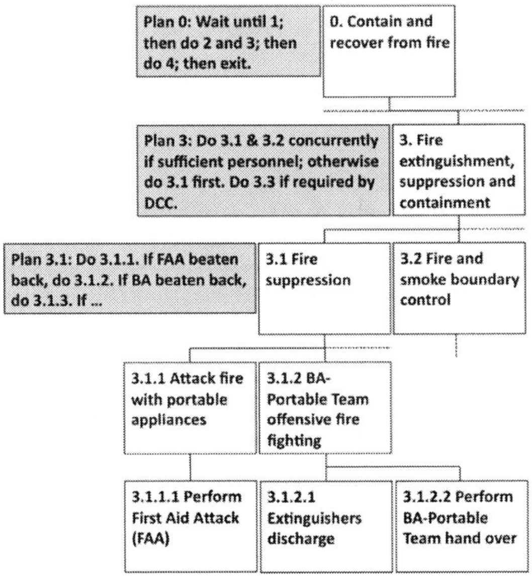

Figure 7: HTA decomposition for fire suppression

The first two levels of HTA decomposition for fire fighting were presented in Figure 3, with a small section of the third level exemplified. Goals 1, 2, 3 and 4 align with the *Manage-Extinguish-Suppress-Contain* process. Each of the sub-goals can be decomposed further, as shown in Figure 7 for Goal 3. Plan 3.1 presents the conditional states for the contingency of various fire suppression goals not being achieved. The aim for Goals 3.1.1, 3.1.2, and others not shown in Figure 7, is to suppress the fire with increasing use of resources. Task 3.1.1.1 Perform First Aid Attack (FAA), for example, is attempted using equipment that is immediately available (such as a portable extinguisher). However, if this is unsuccessful, a team wearing BA, and using fire extinguishers, attempts to suppress the fire – Task 3.1.2.1. Similarly, each of Goals 1, 2 and 4 (not shown in Figure 7) were decomposed to their most primitive tasks. Thus, the HTA decomposition represents an easily understood action plan to perform DC fire recoverability operations, to ensure platform mission effectiveness, platform survivability and crew safety.

4.2 HTA CONSIDERATIONS

When generating the HTA decomposition for the primary DC goals of safely control and recover from fire; safely control and recover from flooding; safely control and recover from toxic hazard incident; and manage action casualty during non-mass casualty incident, respective sub-goals and tasks were created when key words, phrases and descriptions were identified within RAN source material. The challenge was to identify where the goals and tasks were to be located within the HTA decomposition. Some groupings of sub-goals and tasks, or variations thereof, were re-used at various locations within the HTA decomposition. If such commonality did occur, the question then arose 'was it more suitable to separate out the common goal-task hierarchy into one identifiable decomposition located suitably within the hierarchy, or to retain each repeated instance?' This could only be answered through repeated traversal of the HTA decomposition to ensure all relevant combinations of goals and tasks were accounted for within the Plans. For example, Blanket Search may seem to be a common procedure, however its application varies between fire fighting, flood control and toxic hazard response. Therefore, the Plans within the Blanket Search HTA decomposition require conditional statements to ensure the correct tasks are activated according to the incident.

Conditional elements relating to 'if', 'where' and 'when' can cause confusion when identifying the goal, task and Plan hierarchy. Some conditional elements may seem to align easily with a goal, or a task, but should be readily defined within a Plan. For example, a sub-goal description for flood control might be 'Reduce free surface water effect if required'. This description contains the goal 'Reduce free surface water effect', and the conditional element 'if required'. Careful consideration is required and it may result in the need for a new sub-goal with the conditional element being defined within the associated Plan. To overcome this problem also required several iterations and reviews of the HTA decomposition and DC reference material to suitably identify the location of goals, tasks and conditional elements in the HTA decomposition.

Finally, when performing the HTA it was easy to consider just the tasks when trying to identify primary goals and sub-goals. Some goals, sub-goals and tasks, might appear to be the same but the difference between them is: one is the purpose; and the other is the means to achieve the purpose. Forethought needs to given to the separation of tasks from the goals.

The aforementioned considerations are important for the translation of the HTA decomposition to Behaviour Trees for controlling artificial agents representing the crew during M&S. If the HTA decomposition does not accurately represent RAN DC SOPs, the subsequent Behaviour Trees may result in unrepresentative artificial agent behaviour.

5. BEHAVIOUR TREES FOR RAN DAMAGE CONTROL

Behaviour Trees are a popular method in computer games for imbuing Non-Player Characters (NPCs) with artificial intelligence [14, 15, 16]. Instead of assigning an operational state to the NPC, Behaviour Trees assign a behaviour that is to be performed. Behaviour Trees overcome the shortcomings in previous methods for prescribing NPC actions. Behaviour Trees enable scalability, making it easier to include new behaviours; behaviours can be modular, thereby allowing for re-usable behaviour definitions within other NPCs requiring similar behaviours; and Behaviour Trees avoid the exponential increase in transitions between states as the number of states increases [14, 15].

Behaviour Trees are hierarchical in structure, consisting of goals that are decomposed into sub-goals and actions. They can be considered complementary to HTA decomposition, the difference is in their usage: HTA decomposes the behavioural plan; and the Behaviour Tree is the executable plan [14]. The hierarchical nature of Behaviour Trees allows for easy re-use of behaviours across NPC implementations. In the case of M&S for RAN platforms, the NPCs are the intelligent, artificial agents representing the crew and the Behaviour Trees will govern the actions of the artificial agents performing DC operations to recover the platform from the damage incident. The Behaviour Trees, therefore, represent the DC SOPs.

Since the actions of the crew are governed by the DC SOPS, Behaviour Trees are being utilised to control individual crew, and the team behaviour associated with the Repair Bases and DC teams. The use of Behaviour Trees for modelling DCC and C&C is still being reviewed. Functionality for modelling C&C decision-making requires the ability to prioritise goals and to dynamically plan for future possibilities. DCC primarily oversees operations and further understanding of their role is required before implementing their behaviour. Consequently, the HTA decomposition was examined to identify goals and tasks associated with specific levels of the DC organisation. This will facilitate development of behaviour models representing each level of the DC organisation.

Unfortunately, implementing the HTA decomposition as a Behaviour Tree is not a simple one-to-one translation. A Behaviour Tree consists of *Leaf Nodes* known as: Actions – similar to a task in the HTA decomposition; and Conditions – for checking the current state [17]. HTA decomposition has Plans for the specific order of achieving goals, whereas Behaviour Trees have a top-down, left-right implicit priority of performing behaviours. Therefore, the left-most behaviour in the Behaviour Tree will be attempted first. Consequently, translating the RAN DC HTA decompositions requires careful understanding of the HTA Plans and the conditions under which goals are to be achieved.

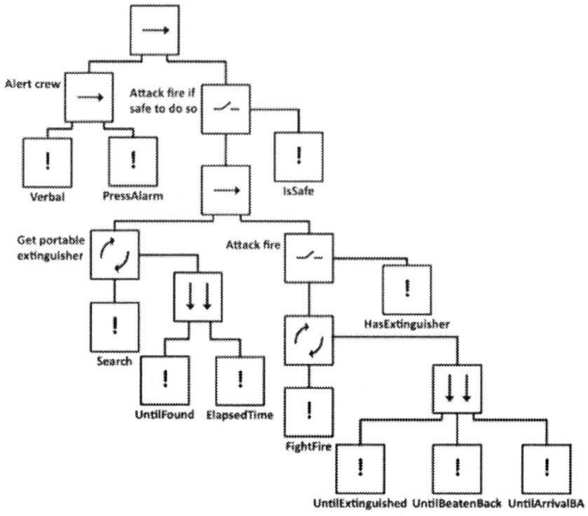

Figure 8: FAA Behaviour Tree

An example Behaviour Tree to perform FAA is shown in Figure 8. This Behaviour Tree represents the amalgamation of Goal 1 and the sub-goal hierarchy associated with Goals 1.1 and 1.2 in Figure 3 with Goal 3.1.1 and Task 3.1.1.1 of the HTA decomposition presented in Figure 7. Comparing Figure 8 with Figures 3 and 7 shows the obvious need for more nodes within the Behaviour Tree to control the conditional activation of behaviours. The rules for the conditional controls are contained within Plans associated with Goals 1, 1.1 and 1.2 (not shown in Figure 3) and Plan 3.1 of Figure 8. Each node within the Behaviour Tree contains a symbol indicating the process it represents, whether it is a task to be performed or a conditional node controlling if the task is to be performed. Table 1 defines the nodes present in the Behaviour Tree shown in Figures 8. Consequently, the behaviour described in Figure 8 indicates that upon detecting a fire, the artificial agent is to verbally raise the alarm and activate any fire alarms. Next, the agent fights the fire, if it is safe to do so, until any one of three states occurs: until the fire is extinguished; until the agent is overpowered by the fire and has to evacuate; or until a fire fighting team wearing BA arrives to take control of fire fighting activities. Other Behaviour Trees will then define actions for the crew agents to perform when they "hear" the announcement of the fire. Software callbacks associated with each node within the Behaviour Tree perform a specific function associated with that node, whether it is to, for example, perform a Boolean conditional check; or activate a simulation of the fire fighting action being performed against a fire model representing the fire encountered. In the latter case, the fire simulation will then determine whether the fire will be extinguished or if it will overpower the fire fighter performing the FAA. Behaviour Trees defining the actions for the crew agents will activate as required, based on the overarching rules governing the crew response as identified in the HTA decompositions.

Table 1: Behaviour Tree node definitions [17]

Node	Definition
!	Action node representing the behaviour to be performed by the agent. It can also represent a condition node to check the current state.
←	Sequence Selector to perform all the child nodes in left to right sequence.
↻	Repeater (more formally known as a Decorator) to repeat the actions of the child node until a defined exit condition occurs.
-/-	Interrupter (also a Decorator) to interrupt the flow of the child nodes when a defined condition occurs.
↓↓	Parallel node to activate child nodes in parallel, exiting when a defined condition occurs.

6. PATH FORWARD

Prior to implementing the HTA decomposition as Behaviour Trees, the HTA decomposition requires validation by experienced RAN DC practitioners. Anecdotal evidence suggests that the HTA decomposition for RAN DC requires refinement, with more granularity in the decomposition. Better granularity will afford more realism from the resultant Behaviour Trees. Then, when validated, the Behaviour Trees will allow the artificial agents to perform their duties autonomously. However, further research is required to understand the critical decision-making process that occurs during DC operations, such as C&C goal prioritisation, and anticipatory planning.

6.1 CRITICAL DECISION METHOD

HTA does not elucidate the cognitive process but it can help identify critical decision points. To understand the cognitive process associated with decision-making during DC, an analysis method known as CDM was suggested – Step 4b in the holistic analysis of crew cognition and interaction during RAN DC scenarios. CDM involves semi-structured interviews with relevant domain experts, whereby the interviewer guides interviewees in recounting a recent incident. The purpose is to gain insight into the decision-making process by asking relevant, probing questions involving numerous themes and constructing timelines for the incident. Key decisions would be identified during the interview and explored during in-depth questioning throughout the interview.

Consequently, the next phase of crew cognitive analysis would be to perform a CDM with RAN DC personnel. Ideally, this would occur immediately after a DC exercise onboard an RAN platform. The aim would be to understand the decision-making process to (primarily) further develop the intelligent artificial agents associated with C&C, DCC and the Repair Bases. Table 2 presents a sample selection of the types of probing themes and associated questions to be utilised in the semi-structured interview for RAN DC.

Table 2: CDM sample questions for RAN DC

Probing Theme	Example Question
Goal Specification	What were your specific goals that guided your decisions and tasks during this phase of the incident response?
Decisions	What decisions did you make during this part of the incident response?
Cue Identification	What did you see, hear and or otherwise sense or intuitively feel that led to your decision/reaction?
Expectancy	What types of decisions were you expected to make at this point during the incident?
Conceptual	In reviewing the timeline, at what points could you have made a different choice/decision? Why?
Influence of Uncertainty	If you were uncertain about the information you were provided, what would have increased your level of confidence?
Information Fusion	If you had multiple information sources for the same information you were seeking, how did you choose which information to use?
Situation Awareness and Assessment	Assume you were performing a handover to another crewmember at that time, what critical information and knowledge would you pass on to help their situation awareness?
Options	Before you arrived at your decisions and performed your tasks, what other options were available to you?
Wisdom of Experience	What common challenges and traps would an inexperienced crew face when making the necessary decisions at this point?
Decision Blocking/Stress	Describe the situation when you found it difficult to process and/or integrate available information during the decision-making process.
Analogy/Prior Experience	What specific training, knowledge or experience led to your reaction or helped you make this decision?
Basis of Choice	Based on your experience, what rules would you devise that might help another crewmember successfully make a similar decision in a similar situation?

6.2 INTEGRATED SURVIVABILITY ANALYSIS

The Behaviour Trees form the basis of recoverability models to be incorporated within ISSAC. Within ISSAC, tools such as mEX [9, 10] and SMARTFIRE (SMF) [18, 19] will respectively model the movement of the crew, and the progression of fire. The recoverability models will receive situation awareness information from mEX and SMF to stimulate crew behaviour as described in the Behaviour Trees. M&S of crew DC behaviour potentially provides the necessary capability within an integrated environment to holistically analyse platform survivability. This is the subject of ongoing research [5] and Behaviour Trees are one possible implementation being examined for modelling artificial agents to respond to DC scenarios likely to be experienced by RAN platforms.

7. CONCLUSION

In a network centric warfare environment, the inclusion of new technologies and an increased level of automation create a changing future that poses unknown and complex challenges for the RAN. Even though 'black swan' events cannot be predicted, ISSAC will enable the exploration of 'what if…?' to analyse the effects of changes to DC SOPs on platform recoverability; to understand crew interaction with autonomous recoverability systems; and to assess RAN platform mission effectiveness, platform survivability and crew safety. Using methods like HTA and Behaviour Trees, crew cognitive analysis and modelling will enable evaluation of RAN platform design philosophies, crew complement and DC tactics to improve platform survivability. The inclusion of recoverability models to simulate crew performing DC SOPs after a damage incident within ISSAC will facilitate the analysis and assessment of configuration changes to the fleet-in-being, and enable trade-off analysis during platform acquisition. ISSAC will thereby enable the quantification of RAN platform survivability in a high threat environment.

8. ACKNOWLEDGEMENTS

The authors acknowledge Director Naval Technical Bureau, RAN, as sponsor of this research. The authors thank Tom Whitehouse and Anthony Travers, Defence Science and Technology Group (DST Group), for support during this research. Thank you to Grant Gamble, lead author [8], for permitting the re-use of the description surrounding *Manage-Extinguish-Suppress-Contain* as presented in [6]. The authors also thank our research partner the Fire Safety Engineering Group (FSEG), University of Greenwich, for their valued contribution towards the development of the ISSAC prototype.

9. REFERENCES

1. DE YONG, L., 'Integrated ship survivability for the Royal Australian Navy', Pacific 2008 International Maritime Conference, Sydney, Australia, 29 – 31 January 2008.

2. LIU, J., and WOOLLEY, A., 'Holistic Analysis of Crew Cognition and Interaction During Naval Damage Control', *Damaged Ship III*, RINA HQ, London, UK, 25 – 26 March 2015.

3. WOOLLEY, A., and GOULD, R., 'An Integrated Capability to Assess RAN Platform Recoverability after a Damage Event', Pacific 2015 International Maritime Conference, Sydney, Australia, 6 – 8 October 2015.

4. *UNITY*, Unity Technologies, Accessed: 1 August 2016, http://unity3d.com/.

5. WOOLLEY, A., EWER, J., LAWRENCE, P., TRAVERS, A., DEERE, S., WHITEHOUSE, T. and GALEA, E., 'A Modelling and Simulation Framework to Assess Naval Platform Integrated Survivability', MAST Europe, Amsterdam, Netherlands, 21 – 23 June 2016.

6. BUCKLAND, M., 'The use of warship vulnerability analysis tools as a measurement of blast and ballistic protection requirements', Pacific 2008 International Maritime Conference, Sydney, Australia, 29 – 31 January 2008.

7. SKEEN, M. and PICKERD, V., *CVAM v2.1 – User Manual*, DSTO-GD-0627, Defence Science and Technology Group, 2011.

8. GAMBLE, G., SUENDERMANN, B., MATHYS, Z. and WOOLLEY, A., 'Fire Modelling and Event Tree Analysis for Naval Platform Fire Incidents', Fire at Sea, RINA HQ, London, UK, 26 – 27 March 2014.

9. GALEA, E. R., LAWRENCE, P. J., FILIPPIDIS, L., BLACKSHIELDS, D., COONEY, D. and SHARP, G., *maritimeEXODUS v5.1 Theory Manual, Revision 1.1*, Fire Safety Engineering Group, University of Greenwich, London, UK, 2016.

10. GALEA, E. R., DEERE, S., BROWN, R. AND FILIPPIDIS, L., 'An Experimental Validation of an Evacuation Model using Data Sets Generated from Two Large Passenger Ships', *Journal of Ship Research* 57(3):155-170, 2013.

11. *Integrated Recoverability Model (IRM)*, Test & Evaluation Solutions, LLC, Accessed: 1 August 2016, http://www.irm-software.org.

12. PAUL S., DANIEL J., NEVILLE S. and WALKER G., 'Hierarchical task analysis vs. cognitive work analysis: comparison of theory, methodology and contribution to system design', *Theoretical Issues in Ergonomics Science*, **11**(6), 504-531, 2010.

13. SHEPHERD, A., *Hierarchical Task Analysis*, Taylor & Francis, London, 2001.

14. DEY, R. and CHILD, C., 'QL-BT: Enhancing Behaviour Tree Design and Implementation with Q-Learning', *2013 IEEE Conference on Computational Intelligence in Games*, Niagara Falls, USA, 11 – 13 August 2013.

15. FLOREZ-PUGA, G., GOMEZ-MARTIN, M., GOMEZ-MARTIN, P., DIAZ-AGUDO, B. and GONZALEZ-CALERO, P., 'Query-Enabled Behaviour Trees', IEEE Transactions on Computational Intelligence and AI in Games **1**(4), 2009.

16. ISLA, D., 'Handling Complexity in the Halo 2 AI', Game Developers Conference, 2005.

17. JOHANSSON, A. and DELL'ACQUA, P., 'Comparing Behavior Trees and Emotional Behavior Networks for NPCs', The 17th International Conference on Computer Games: AI, Animation, Mobile, Interactive Multimedia, Educational & Serious Games, Louisville, USA, 30 July – 1 August 2012.

18. EWER, J., JIA, F., GRANDISON, A., FROST, I., GALEA, E. R., and PATEL, M. K., *Technical Reference Manual and User Guide for SMARTFIRE 4.3*, Fire Safety Engineering Group, University of Greenwich, London, UK, 2013.

19. WANG, Z., JIA F., GALEA, E. R. and EWER, J., 'A CFD Simulation of a Post-Crash Aircraft Fire Test', *Journal of Aircraft* **50**(1):164-175, 2013.

10. AUTHORS BIOGRAPHY

Anthony Woolley is a defence scientist in Vulnerability, DC and Recoverability (VDCR), DST Group. His responsibilities include co-ordinating development and implementing elements of ISSAC. His previous experience includes platform configuration management M&S; crew operational relief analysis; submarine systems analysis; and platform systems integration and habitability analysis. Anthony has a Bachelor of Science (Honours) in Computer Science and Physics from the University of Tasmania.

Jenny Liu joined DST Group in 2011 as an Executive Support Officer. She has a Bachelor of Arts in Psychology from The University of Melbourne. During six months of 2015, Jenny worked full-time in VDCR as a defence scientist to gain domain specific research experience. Jenny's research has included RAN cognitive task analysis to understand the crew component of RAN Platform Recoverability and DC responses. Jenny also facilitates the implementation of the DST Strategic Plan 2013-2018; project-managed DST Group's Leadership, Accountability and Performance Improvement Initiative; and is involved in the development and implementation of the Next Generation Technology Program. Jenny is also responsible for facilitating the review and improvement of the internal management for the DST Group/Defence Materials Technology Centre program.

SHIP INTERNAL AND SHIP EXTERNAL EFFORTS TO REDUCE THE HUMAN FAILURE IN SHIP OPERATION

E Aa Dahle, Dr. Ing., Consultant, Norway. **K Bentzrod,** Master Mariner, Hoegh LNG, Norway, **N H Lund,** Master Mariner and Maritime Government Pilot (Ret.), Norway

SUMMARY

The dominant part of ship accidents is constituted by powered grounding, in cost if not in loss of lives. Groundings are mainly caused by human failure. Over the years, numerous efforts have been made to reduce or eliminate human failures. From the point of view of risk analyses, risk reduction could be attained by reduction of frequency of human failures, and/or by reduction of grounding consequences. The latter is not a subject of this paper.

NOMENCLATURE

AIS.	Automatic Identification System. AIS. The system is mandatory for ships above 45m in length, or with tonnage above 300GRT. The system submits information to other ships, and also to VTS. The info consists of i.e. position, course and speed over the ground (E and I)*
AToN.	Aids to Navigation. Physical means, or virtual by AIS (E)*
BNWAS .	Bridge Navigation Watch Alarm System. The system is mandatory for ships above 250BRT. If the OOW does not respond, alarm is submitted to a manned station onboard, i.e. a "dead man button", often based upon immobility on the bridge. Also leisure crafts may acquire a relatively simple AIS (I)*
COLREGS.	The International Regulations for Preventing Collisions at Sea. International rules of conduct for commercial ships. Includes rules for lights and signals, and TSS (E)
DSC.	Direct Selective Call. This is a radio signal sent to one, or to a group of ships from a VTS. Part of GMDSS. If no response, an alarm could be sent to a manned station on board. Long range (E)*
ECDIS	Electronic Chart and Display System. The system may be fully implemented from 2018. The screen-based system will display relevant electronic charts with selected data from radar, satellite systems (up to 4 are available) Guidelines in SOLAS (E and I) *
E-navigation.	A ship with ECDIS must i.a. submit its sailing plan for approval to the relevant VTS. VTS will approve (or not approve) the plan, and monitor that the vessel is following the plan accordingly. Under development in IMO (E and I*)
FSA.	Formal Safety Assessment. A method of cost/benefit analysis based upon risk analysis. Guidelines by IMO (-)
GMDSS.	Global Maritime Distress and Safety System. Part of, and includes DSC, that can alert an errant ship (E)*
IALA.	International Association of Lighthouse Authorities. An international system of buoys used globally for navigation (-)
IMO.	International Maritime Organization. Under UN, IMO work out conventions that are mandatory for nations that sign the convention. Coastal states have jurisdiction over national waters (-)
OOW.	Officer On Watch (I)*
RINA.	Royal Institution of Naval Arcitects
SOLAS.	The International Convention for the Safety of Life at Sea. An international maritime treaty obliging signatory states to minimum standards in construction, equipment and operation (E and I)
STCW.	The International Convention on Standards of Training, Certification and Watch keeping for Seafarers. Specifying qualification requirement for all on board maritime personnel. Guidelines in SOLAS (E)
VHF.	Very High Frequency radio, used on most ships and vessels. For coastal areas (E and I)
TSS.	Traffic Separation Scheme. Such systems will define a sector, normally one for each direction that shall be followed under normal condition. Smaller ships may be assigned a limited, outlaying sector. Guidelines in COLREGS (E)*
VTS	Vessel Traffic Service, similar to air traffic control for aircraft and road administrations. In Norway there are 5 VTS, which survey some particularly sensitive sea areas along the coast. VTS use i.a. VHF, AIS and radar. Guidelines in SOLAS (E and I)*

1. INTRODUCTION

The dominant part of ship accidents is constituted by powered grounding, in cost if not in loss of lives. Groundings are mainly caused by human failure. Over the years, numerous efforts have been made to reduce or eliminate human failures. From the point of view of risk analyses, risk reduction could be attained by reduction of frequency of human failures, or by reduction of grounding consequences. The latter is not a subject of the paper.

Reduction of grounding frequency can be achieved by ship internal efforts, ship external efforts or by combinations of both. Significant efforts are listed in the Nomenclature above. `E` signifies External efforts, while `I` signifies Internal efforts. The asterix * signifies that the efforts will be briefly discussed in this paper.

So far, efforts introduced since the 1980-s seem to neither have been coordinated nor evaluated. Hence, there is an almost random implementation of partly overlapping efforts that have not so far resulted in a significant reduction of grounding frequency.

Ship internal efforts are mostly worked out through IMO, while ship external efforts such as AToN are worked out by i.e. IALA. It seems that pre as well as post evaluations of such effort are weak, for separate as well as for coordinated efforts. A typical example of this situation is `navigator fatigue`. A ship internal measure is e.g. BNWAS, while a ship external measure is warning an errant ship by VTS. These efforts are overlapping, and have not been evaluated in combination. In the paper, a coarse assessment method of impact of ship internal, ship external and combined efforts on powered frequency caused by human failure is presented.

2. MODELLING OF POWERED GROUNDING

In risk analysis, it is common to model the grounding per year and for a specific area by [2],

$F = N \cdot (p_{Generic.} \cdot p_{Geometric.} \cdot p_{Efforts})$

N = number of ships passing the area per year (N/2 each way)

$p_{Generic}$ = historical, `generic` probability for a skip to proceed on a `dangerous` course. A common figure used in several reports is 2·E-4 figure (one of 5000) is based on the assumption that the OOW, in coastal waters checks the position approximately every 20 minute

$p_{Geometric}$ = probability of being on a grounding course against a fixed object. This figure is based upon a lateral ship distribution. AIS has made it possible to assess the ship distribution as e.g. in [3]. A (slightly adjusted) Normal distribution is widely used in later years, with a standard deviation of about 1/3 of the half with of the navigable fairway. Hence, $p_{Geometric} \approx f(x) \cdot b$

f(x) = the functional value of the normal distribution at the distance x from the middle of the fairway

b = lateral breadth of the object

$p_{Efforts}$ = probability that `ship internal efforts` and/or `external efforts` influence on the grounding caused by human failure

In this paper, only the approximately 60% of powered grounding frequency caused by human failure is considered. The assumption that the efforts in question are fully implemented on board does not take into account that this may not be the case.

3. HUMAN FAILURE

A considerable fraction of grounding is caused by what is termed `human failure`. In presenting statistics in Norwegian waters from 1997 to 2014, the `human failure` fraction is given as 0.6-0.7. The remaining fraction is caused by `technical failures', 'severe weather` and `other causes`. An interesting result from the statistics is that the frequency of grounding in general (per sailed distance), as well as the fraction of `human failure`, are rather constant. This is in spite of numerous efforts. Typically, ship internal measures and ship external measures are administered by separate national bodies. Isolated efforts are sometimes assessed by FSA, often restricted to new technical installations, leaving out the potential for human failure.

4. EVENT TREE (ET) MODELING OF EXTERNAL AND INTERNAL RISK REDUCING EFFORTS

One disturbing observation can be made without using an Event Tree:

- With BNWAS fully implemented, the fraction of powered grounding caused by fatigue and other incapacitation modes would be near zero. This is evident from the equation above.

Could it be that the BNWAS is often `sabotaged`? For the motion type by e.g. hanging CDs dangling before he sensor (motion type), or by simply disconnecting the dead man button type? Or is the button routinely pushed without looking into radars nor out through windows? The purpose of such `sabotage` or negligence could be due to pending paper work such as chart corrections. Nevertheless, mitigation of fatigue is still a matter of concern at IMO. The matter is often connected to number of OOW to be one or two, and hence associated with manning costs.

Could it be that BNWAS, by neglecting the human element, is a failure?

The ET is intended as a simple means to indicate the order of magnitude of effect of ship internal and ship external in conjunction. It is necessary to assign coarse, rather subjective figures to the branch probabilities in the ET. Nevertheless, the ET provides a tool that is simple to use, and which clearly exhibits numerical figures, based on the assessments made.

It is noted that the efforts implemented are assessed to be only partly implemented. Hence the modeling illustrates that even with the significant strengthening of VTS authority as indicated later in the paper, parts of the human failure fraction will remain.

From data in [2], the often used generic factor 2E-4 is considered as valid in the1980-s. For further simplification, human factor fraction p_{1980} is set to unity. Powered grounding is restricted to *one* ship always on a `dangerous` course against a fixed object, i.e. $p_{Geometric} = 1$. For assessment of $p_{Efforts\ since\ 1980}$, an ET has been made. Reference is made to Figure 1, 2, 3 and 4 at the end of the paper.

The following efforts, introduced since the 1980-s, are considered in the ET as shown in Table 1.

Table 1: Assessed reduction in human failure contribution for powered grounding.

Situation	Efforts	Reduction in human failure
Figure 1. As today (2016). One OOW under normal conditions.	-	15%
Figure 2. As today (2016). Two officers on the bridge, ore one OOW and one pilot. under normal conditions.	-	81%
Figure 3 After implementation of E-navigation (2018).	ECDIS, with approved sailing plan exhibited on board and at the VTS. AIS integrated in ECDIS.	85%
Figure 4 Additional efforts after implementation of E-navigation. (After 2018).	AToN supplied with virtual or physical AIS (shown on the ECDIS screen). BNWAS, transferred to VTS after no response to on board alarm. Thereafter DSC warning from VTS (after time lag, transferred to manned location onboard, thereafter GA through DSC). Off course alarm on board and at the VTS. If in conjunction with no BNWAS or VHF response, GA through DSC. VTS given strengthened authority to give navigational order to errant ships, for holding back ships in dangerous crossing of TSS (using AIS and/or radar simulation), for overtaking as well as in narrow parts of TSS fairways (e.g. under bridges), for all important coastal fairways	91%

Table 1 Reduction in human failure part of powered grounding frequency caused by external and internal efforts

5. CONCLUSIONS

Numerous ship internal and ship external efforts to reduce the frequency of powered grounding caused by human failure have been implemented since the 1980-ties. From statistics, the result is very disappointing as the powered grounding frequency over the years has been almost constant.

- In the simple ET assessment presented, a future 91% reduction of human failure could be expected, provided that TSS is implemented in coastal fairways. Without TSS it is difficult for VTS to define dangerous deviations from the course line.
- VTS is given the authority to release GA onboard when TSS is transgressed (without prior info to VTS)

- A much stricter VTS control regime could be considered. This aspect may be implemented in the on-going E-navigation development process.

However, changes in related laws may be required, and sailing regulations must be developed. Pilotage from VTS, as well as exemption from use of pilot onboard might be considered in conjunction with Pilot exemption certificates. Sailing regulations related to size of ships, speed, strengthening of bridge manning, as well as environmental condition should be taken into account.

6. REFERENCES

1. The Coastal Administration of Norway. `The safety analysis of groundings and collisions in Norwegian waters 2014`. In Norwegian. *Report No. 1908Z31-6. 2015*

2. KRISTIANSEN, S., `Safety Management and Risk Analysis`, *Elsevier* 2004.

3. DAHLE, E. et al Risk Analysis of Gåsøyrenna (Oslo fjord). In Norwegian. *Safetec Report 2012-05-30.*

7. AUTHORS BIOGRAPHY

Dr. Ing., **E Aa Dahle,** FRINA, Marine and maritime consultant. He has been professor at the Technical University of Norway, and has i.a. worked with DNV GL. He holds a Class 3 DOC.

Capt. **K Bentzrod**, Master Mariner, presently at Hoegh LNG. He has been a master of ships for more than 30 years, and has participated in development projects for the Coastal Administration of Norway.

Capt. **N H Lund**, Master Mariner and Maritime Pilot (Ret.). He holds a Master`s Certificate, and has practiced as a Maritime Government Pilot in Norwegian waters for more than 35 years. He has participated in numerous VTS and other development projects for the Coastal Administration of Norway.

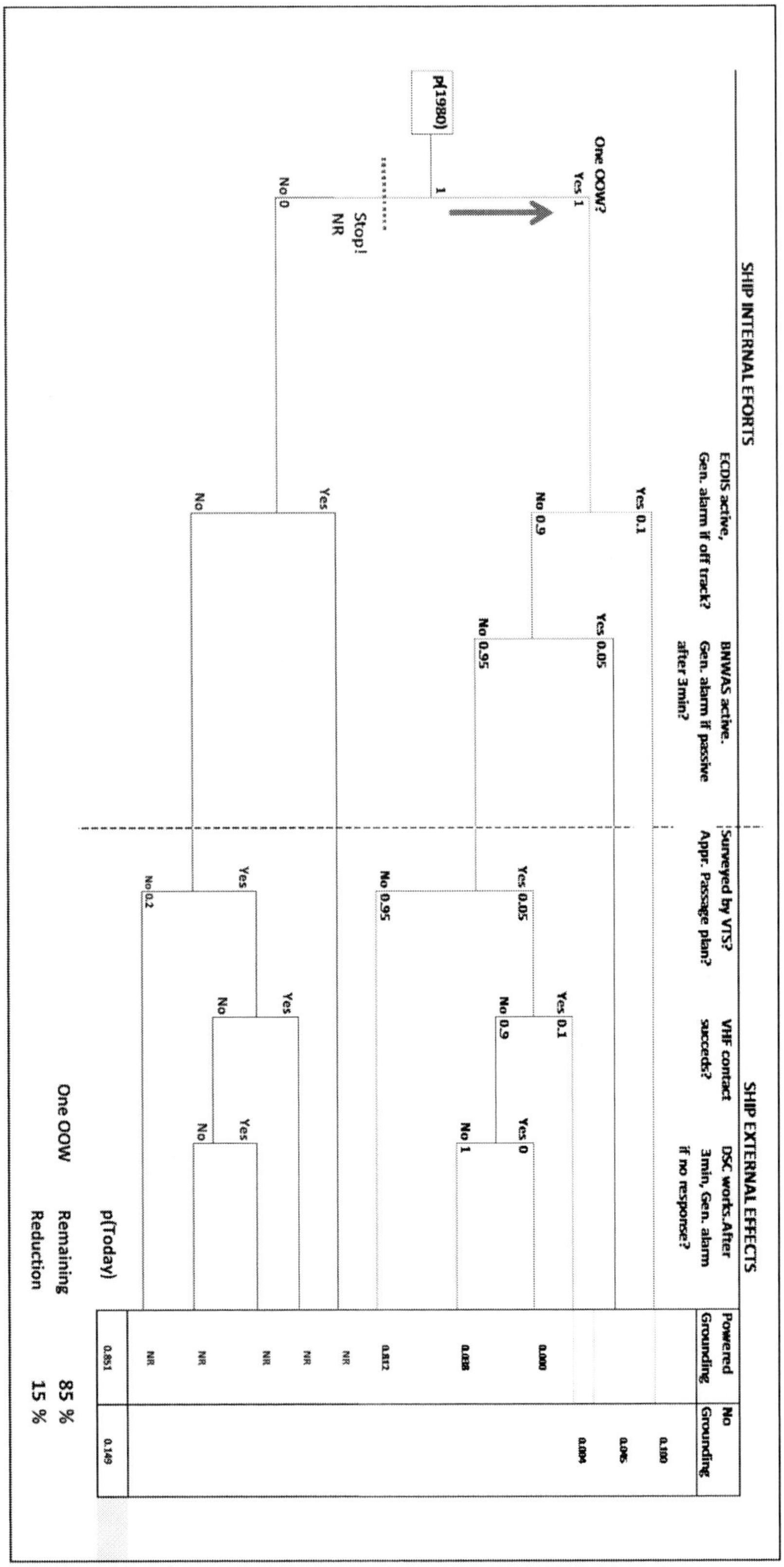

Figure 1: As today (2016). One OOW under normal conditions

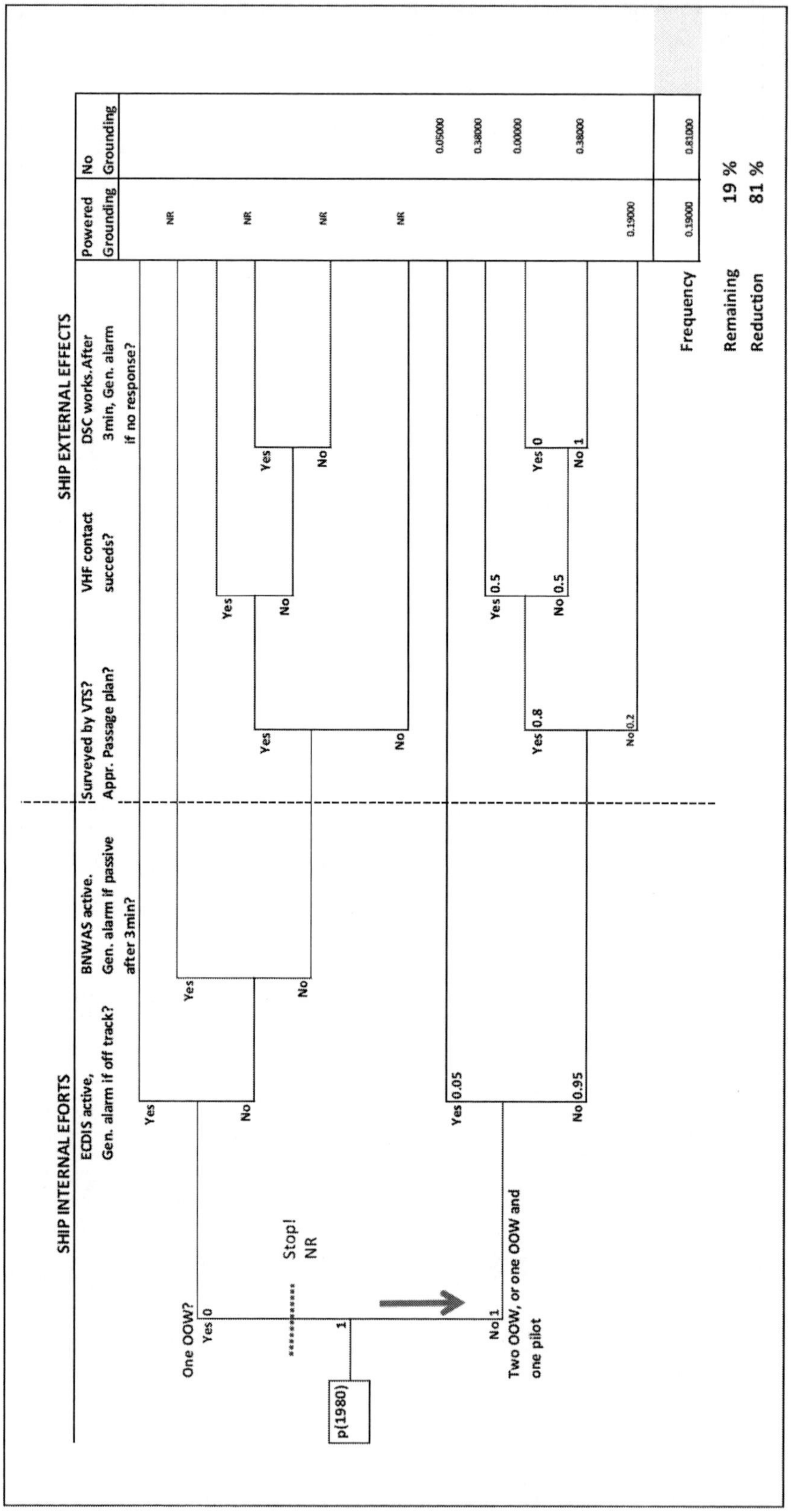

Figure 2: As today (2016). Two officers on the bridge, ore one OOW and one pilot.

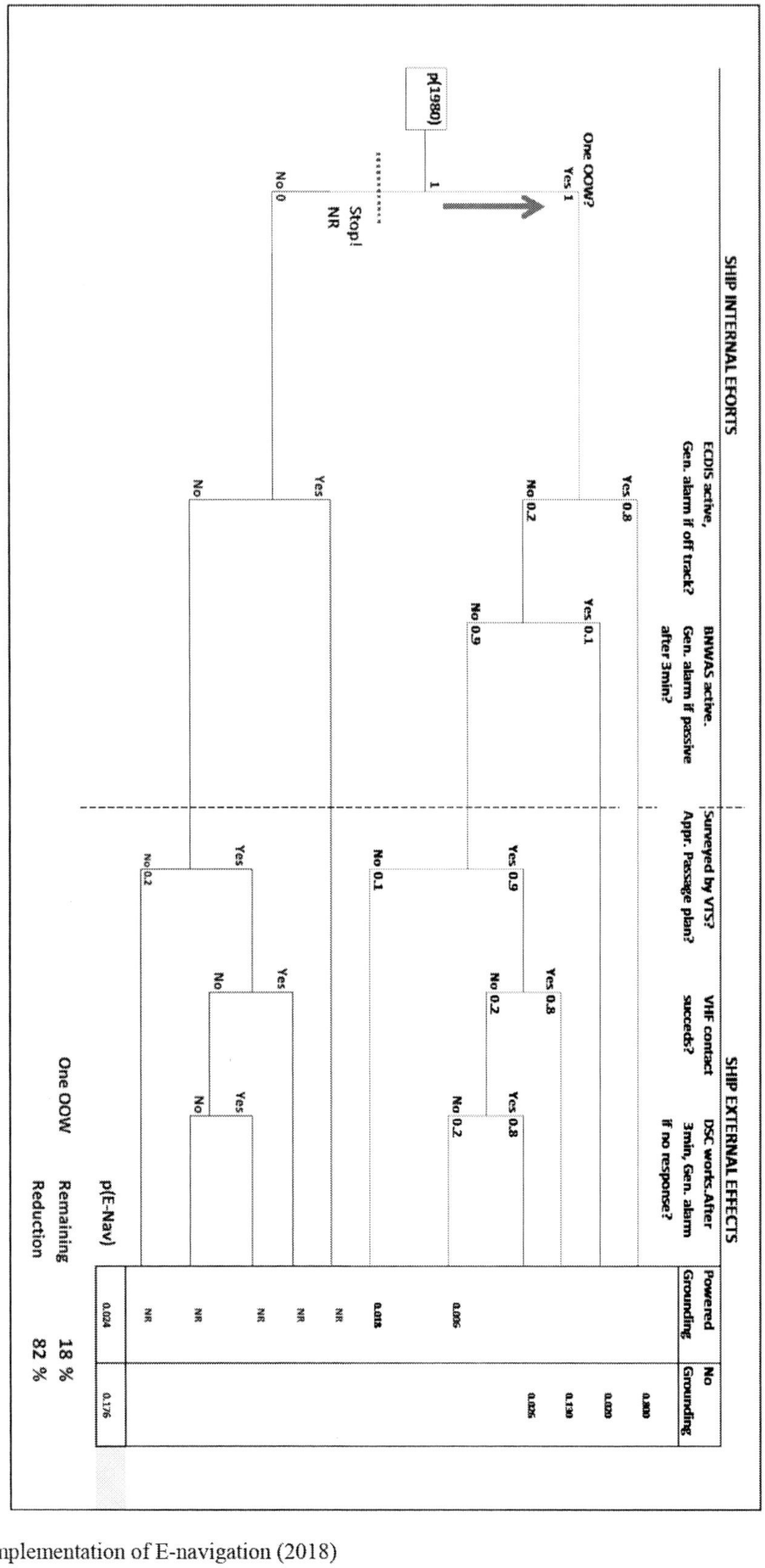

Figure 3: After implementation of E-navigation (2018)

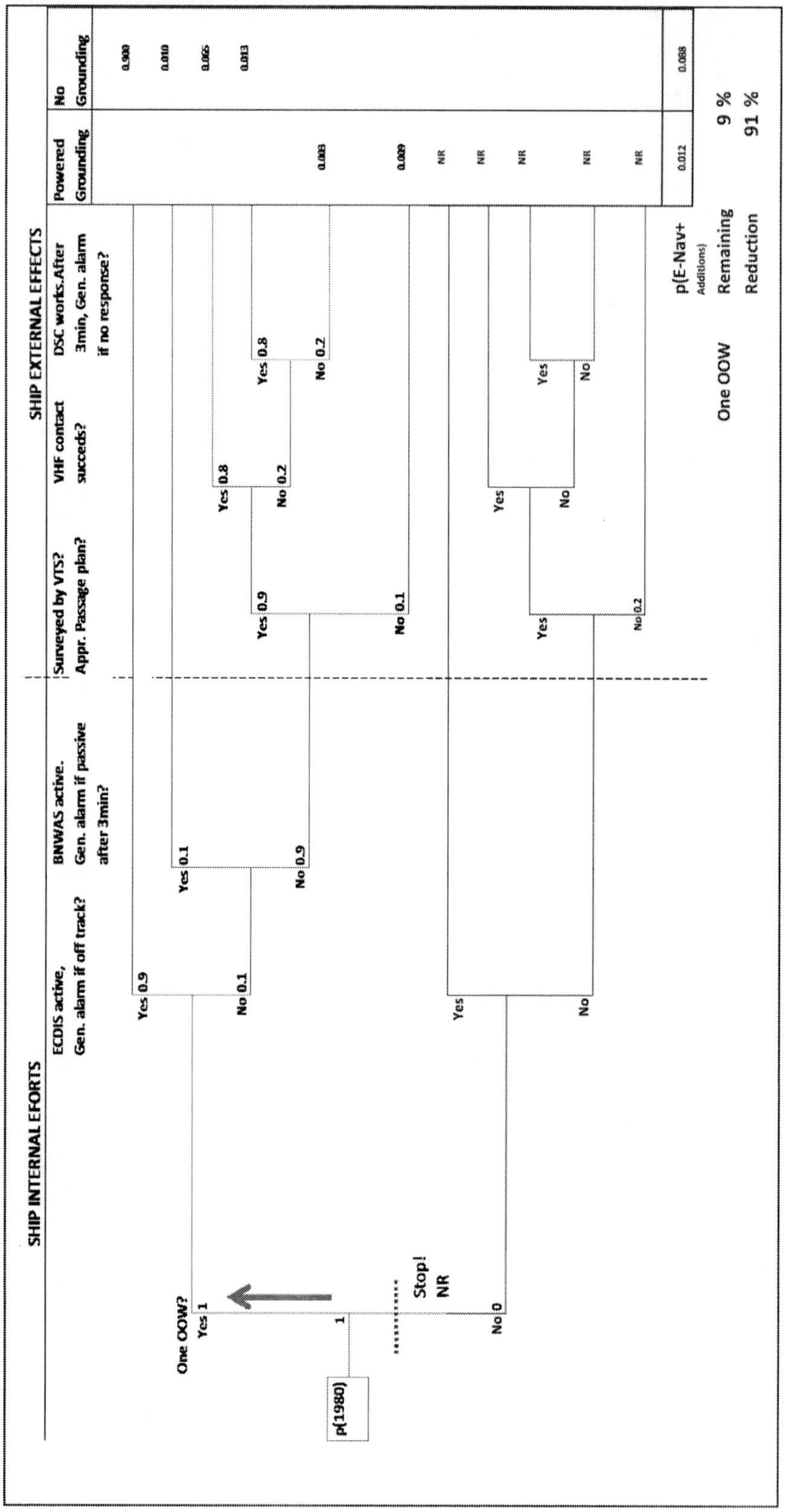

Figure 4: Additional efforts after implementation of E-navigation. (After 2018)

HUMAN CENTERED DESIGN FOR AN INNOVATIVE TRAINING SYSTEM USING VIRTUAL REALITY: NAVAL HELOHANDLING TRAINER (NHT).

C Maïs, DCNS Research, France – L **Lemonnier**, CLARTE, France – **Y Bouju,** DCNS, France and **A Bouchet**, CLARTE, France.

SUMMARY

This paper presents the Human Centered approach used for an innovative training system using a virtual reality (NHT) headset (Oculus RIFT). This system aims to train flight-deck crew (HCO, LSO...) on naval ships. Firstly, the importance of NHT as an asset for training of such crews and a modern naval force is discussed and the user requirements it fulfills are described. Details concerning the system developed by CLARTE and DCNS are then given in the second section. This paper presents the Human Centered approach used to design the NHT solution for the flight-deck director. This solution is based on a Virtual Reality Environment using a head-mounted display and a real-time Gesture Recognition toolkit. One of the main challenges is to develop a system that is sufficiently representative of real-life conditions (ship platform movements especially) while not causing virtual-reality sickness during training sessions. In a third part, the evaluation of this solution is presented, especially the "Virtual Sickness" aspect.
Finally, the results are discussed.

NOMENCLATURE

HCO	Helicopter Control Officer
LSO	Landing Signal Officer
NHT	Naval Helohandling Training system
OOD	Officer Of the Deck

1. INTRODUCTION

Sea trials with helicopter maneuvers are increasingly rare and result in considerable cost. As a result, team members in charge of helicopter maneuvers receive very short training to learn the best practices and the correct procedures.DCNS wanted to develop a new educational tool that was both modern and attractive for the benefit of helicopter platform teams engaged in ship warfare.

This system, called NHT (**NAVAL HELOHANDLING TRAINER),** is a multi-player simulator that aims to educate and train the aviation teams for the deployment of helicopters on naval ships under realistic conditions. NHT creates many opportunities for training scenarios including cases of non-compliance that can generate a hazard.

NHT preserves the potential of vessels and helicopters. This simulator has once again demonstrated the potential of virtual reality simulators associated with motion capture for teaching and training. The possible evolution of technology components and crew scenarios is an asset for this simulator. Virtual reality systems allow learning by doing, as in the real world [1].

After a description of the NHT users and the system itself, this paper focuses on the human-factor evaluation of virtual-reality sickness. NHT received the LAVAL VIRTUAL Award in 2016.

2. NHT USERS

This part describes the FLIGHT-DECK CREW: HCO and LSO [2]. Oher users, such as the instructor or safety crew, are not described.

HCO
The HCO exercises overall control of flight operations by ensuring that the cutter, flight deck, and helicopter are ready for each flight phase.
The HCO keeps the OOD informed of flight-operation requirements so that the cutter may be maneuvered as required to meet the needs of the helicopter and safe navigation.

LSO
LSO is in charge of the flight deck during ship-helo operations.
The LSO and pilot hand signals are the primary method of communication between the LSO and the helicopter during all flight-deck movements. Light wands are used at night (see Fig. 1).

Tiedown Crew
The tiedown operation is critical to a safe helicopter landing and requires a coordinated and rapid team effort.

Rescue Crew
Personnel selected for flight-deck rescue crew duty must be capable of operating in adverse conditions. They may be required to extract unconscious personnel from inside the helicopter or perform other duties which require above-average physical strength and endurance.

Figure 1: LSO on a naval ship

3. NAVAL HELOHANDLING TRAINING SYSTEM

NHT system has been designed through a Human Centered approach.

- User needs: To complete previous task analyses, interviews were performed with LSOs to identify their needs (visibility, communications, etc.) during takeoff and landing on naval ships. Interviews are also focused on the difficulties generated by the environment: wind, platform instability, etc. This step allows us to define the right level of virtual simulation (deck movement depending on weather conditions for example).

- NHT complies with Ergonomics recommendations for Virtual Reality environments for avatar uses, representativeness, and feedback [3].

According to these principles, CLARTE has designed and implemented the NHT system.

NHT has 4 areas:

- **Observer area**
  This area allows the observation of different views on TV screens: LSO, HCO, virtual simulation…

- **Instructor area**
  This is a 2-screen workstation.
  The instructor can define and select the scenario, see what the LSO sees and does, and analyze his performance (correct arm position).
  The helo pilot communications are also simulated by the instructor.

- **Virtual Aviation room**
  The HCO has a screen simulating the HCO dashboard and another screen simulating the external watch on flight deck and environment. He also has a touchscreen simulating green and red light commands and radio.

- **Virtual Flight deck area**
  This area allows LSO immersion (see Fig. 2).
  This Platform is fixed-base.
  This is an area of 3x3 meters with 12 optical cameras.
  VICON, 1 PC station, 1 Oculus Rift DK2 mask, and a capture motion suit.

Figure 2: LSO in the NHT system

Sensors placed on the body allow the definition of a virtual avatar. The LSO performs arms movements to indicate to the pilot the required action (forward, backward, up, down, etc.) according to the normal takeoff or landing sequence. He can also communicate with others users.

Other users (tiedown crew) are simulated with avatars. They simulate normal actions.

NHT architecture is represented below (see Fig. 3).

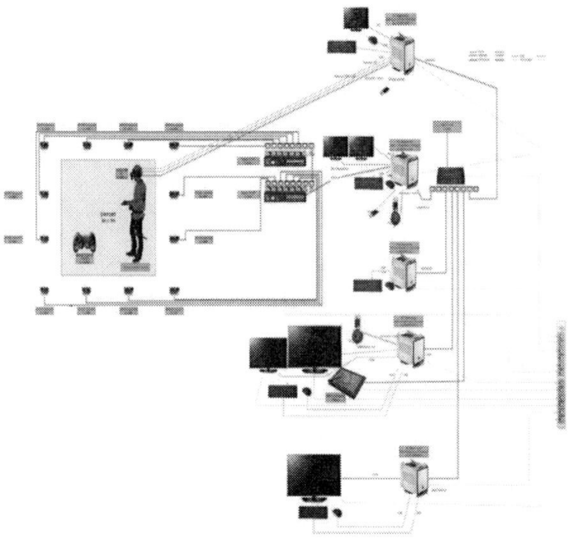

Figure 3: NHT architecture

4. HUMAN FACTOR EVALUATION

The HF evaluation was conducted in 2 steps: The first step deals with utility and acceptability. This is a necessary step, without which the project would stop.
Reviews were performed with marine staff users to present NHT concepts and systems. This is a necessary step but not sufficient in itself. The usability of this system must be evaluated. One major issue with virtual systems is virtual sickness. Indeed, it is not possible to have a system which causes virtual sickness when used in representative situations (training sessions).

4.1 VIRTUAL SICKNESS

VR sickness represents a barrier for many users of this immersive technology. People who experience VR sickness will often stop using their headsets. Some users of virtual environments experience adverse effects known as simulator sickness. Common symptoms are generally grouped into nausea, oculomotor discomfort, and disorientation. Another indicator of simulator sickness is the difference in postural stability before and after a simulator experience. Virtual sickness can occur and remain after the immersive session and cause difficulties once back in the real-life context.

Many factors are important for VR sickness [3]. We focus on relevant factors considering NHT characteristics, situations and user characteristics.

Some factors are simulator related: In a fixed-base simulator, the visual system senses motion while the operator remains stationary. This conflict of sensory cues may result in simulator sickness that manifests itself in terms of a subject becoming nauseous and/or disorientated. Because NHT is a fixed-base system cue conflict can occur and generate VR sickness. Other factors relate to person-specific characteristics such as age, gender, experience.

- Age:
  - Young people are more susceptible to virtual sickness.
  - LSOs are often around 40 years old. But if NHT is extended to tiedown crew, it will need to be extended to younger people.
- Gender:
  - Women tend to be more susceptible to motion sickness: it was reported [4] that females had higher post-exposure oculomotor discomfort scores (20.84) than males (11.37).
  - Although LSOs are often male, this job is open to females.
- Experience with the real-world task:
  - Pilots with more flight experience and little simulator experience are more prone to simulator sickness.
  - It is necessary that the training and tutoring system be suitable for both inexperienced and experienced persons.

Some factors are also situation and task related, such as duration and standing posture.

- It has been found that intensity and duration of ataxia increases with duration. Previous research has found that a twenty minute exposure to VREs can increase sickness symptoms in over 60% of participants [4, 5].

- Seated persons are less prone to sickness because of reduced demands on postural control with respect to standing persons.

Taking into account all these factors and NHT characteristics (fixed-based system, standing posture, etc.), it was necessary to investigate and evaluate NHT from the virtual sickness point of view.

4.2 METHOD

Different hypotheses must be tested:

- H1: No virtual sickness when using NHT in all situations (good or bad weather with deck movements) and for all types of person (age, experience with the real-world task, or gender).

- H2: No virtual sickness as a function of duration (many training sessions can be performed successively).

4 sessions were performed successively (always in the same order):

- Session 1: good weather with instructor avatar watching arm movement for each action.

- Session 2: Good weather without instructor avatar.
- Session 3: Bad weather with instructor avatar.
- Session 4: Bad weather without instructor avatar.

The evaluation scenario involves 8 steps. With the instructor avatar, the instructor shows all gestures at each step and the LSO apprentice has to repeat the gestures (see Fig. 4). The different steps are starting engine and rotor, removing chocks, taking off, landing, insert chocks and stop the rotor and the engine. The apprentice performs all gestures (see Fig. 5) and answers the questions after each scenario.

Figure 4: Scenario steps

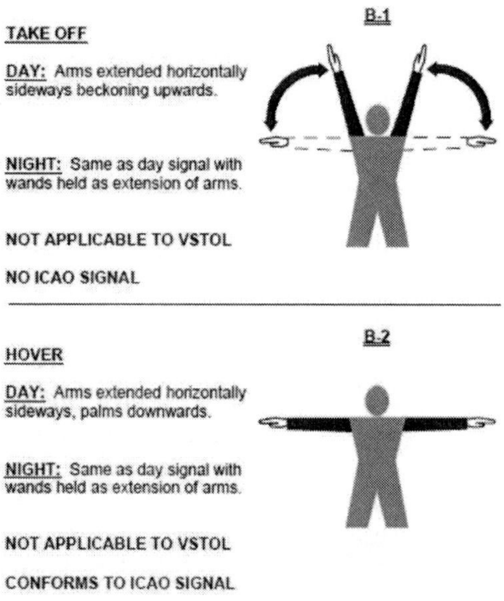

Figure 5: Gesture examples

Each session lasts between 7 and 15 minutes, representative of training sessions. Of course, real take-off and landing operations are longer. Before the first session and after each session, apprentices complete the SSQ questionnaire and perform the ROMBERG Test [5].

4.3 RESULTS

User' profiles

Because it is difficult to have a large panel, representative of future navy users, other persons are invited to participate.

There were 10 subjects with following profiles:

- Gender: 2 female and 8 male.
- Experience with the real tasks: Only 2 subjects had experience as LSO. The other subjects were students.
- Age: 4 users were between 35 and 45 years old (representative of LSOs). The other subjects are younger (between 20 and 30).

Virtual sickness

The graph below (Fig. 6) shows SSQ global, nausea and oculomotor discomfort means per session.

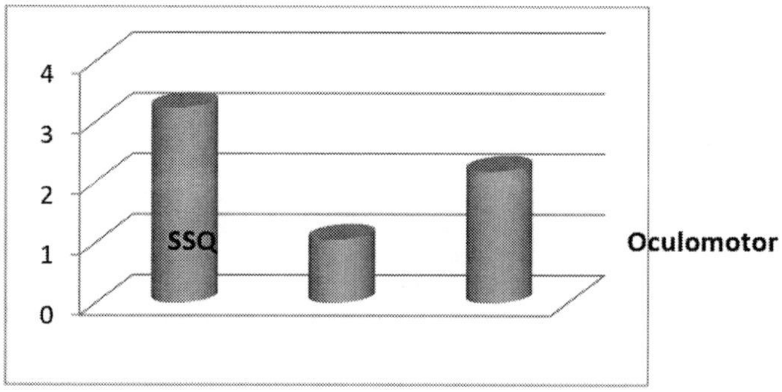

Figure 6: SSQ score (mean)

SSQ global as well as nausea and oculomotor discomfort scores can be considered as "nothing at all" because they are under the lower quarter. Even if oculomotor discomfort is more important than nausea, it can be considered as insignificant.

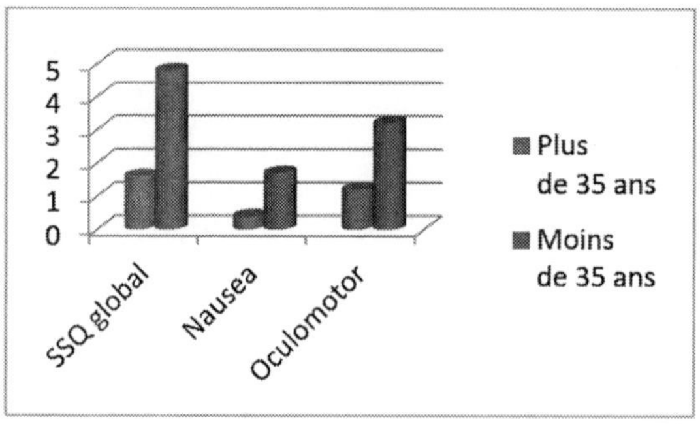

Figure 7: Age effect

Results confirm that younger people are more sensitive to virtual sickness even if it is insignificant, for each dimension of virtual sickness. Gender effects can't be analyzed because of small size of female people. Furthermore, the women were young. It is the same for experienced people.

Weather effects are not significant as shown in figure 8.

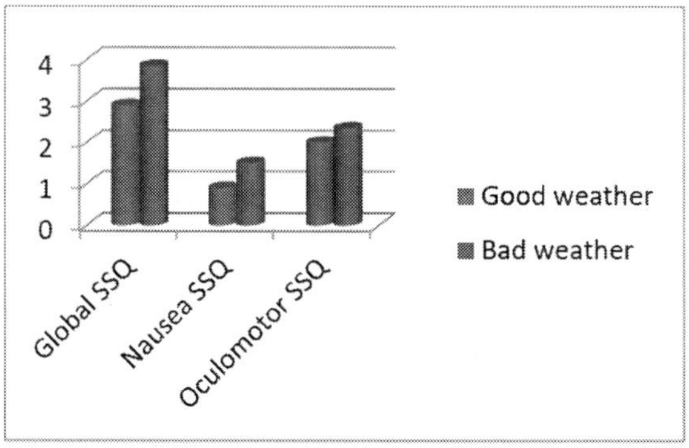

Figure 8: Effects of weather conditions

Duration effects on virtual sickness

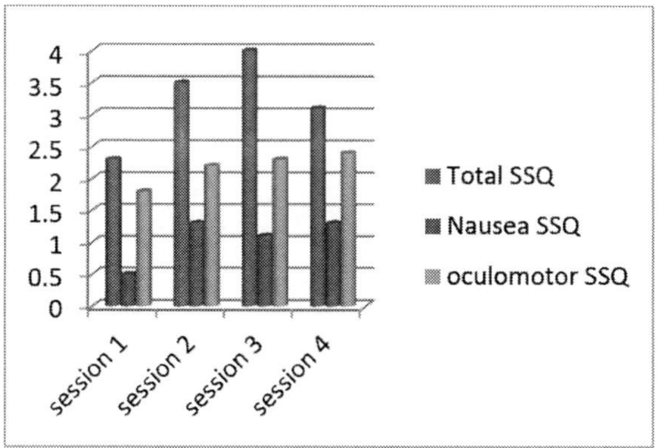

Figure 9: SSQ score (mean) by session

We notice (fig 9):

- An increase for the global SSQ mean according to the sessions except for the last session.
- Oculomotor score are more important for all sessions than nausea score.
- Nausea score seems to be independent to duration.

Comparatively to previous results with head mounted display (see fig.10 cited by [4]), no virtual sickness has been observed with NHT.

Study	Nausea	Oculomotor Discomfot
Kolasinski & al. (1998)	18,13	16,11
Ehrlich & al. (1998)	14,91	15,4
Stanney & Kennesy (1998)	30,21	25,74

Figure 10: SSQ score with HMD

ROMBERG test allowed consolidating the previous data by exploring stability (observations completed feelings). The essential features of the test are as follows: The subject stands with feet together, eyes open and hands by the sides. Then the subject closes the eyes while the examiner observes for a full minute. We can observe very small body movements (not significant) with immersive situations and no significant increase with duration and weather conditions.

5. CONCLUSIONS

Many conclusions in different area can be made:

- Like every simulator, NHT allows a team to be trained in very close to real conditions. Due to its immersive character, virtual reality, which does not require any major infrastructure, can be used to rapidly address different fields and situations, i.e., different vessels and different operational conditions, from the simplest (good weather) to the most difficult (boundary conditions up to the occurrence of an accident), which is not feasible in real life or under real training conditions.
- New technologies less intrusive for motion capture can be added.
- A Human Centered design approach allows the definition of the NHT system, suited to user needs.
- The evaluation of the NHT system allowed us to ensure that the future product would not generate virtual sickness in similar "real conditions". With a duration exposure about 30 minutes, users of the sample did not experience sickness.

6. ACKNOWLEDGEMENTS

First, we would like to thank the users and customers, especially the French Navy, who participated in the NHT definition and evaluation phases. We would also like to thank the DCNS NHT project crew, which placed its trust in CLARTE and HF and the virtual DCNS team.

7. REFERENCES

1. BURKHARDT, J.-M., "Réalité virtuelle et ergonomie : quelques apports réciproques", *Le travail humain*, Vol. 66, p. 65-91, 2003.

2. SHIPBOARD-HELICOPTER OPERATIONAL PROCEDURES MANUAL

3. KOLASINSKI, E.M., "Simulator sickness in Virtual Environments", *Technical reports US Army 1027*, 1995.

4. MOURANT, R.R. and THATTACHERRY, T.R., "Simulator sickness in a virtual environment driving simulator", *Proceedings of the IEA 2000/HFES 2000 congress*, 2000.

5. KENNEDY, R.S., LANE, N.E., BERBAUM, K.S., & LILIENTHAL, M.G., "Simulator Sickness Questionnaire: An enhanced method for quantifying simulator sickness." *International Journal of Aviation Psychology, 3*(3), 203-220, 1993.

8. AUTHORS' BIOGRAPHY

Alexandre BOUCHET currently holds the position of research & development manager at CLARTE. He is responsible for specifying and leading the virtual- and augmented-reality research strategy of CLARTE. His previous experience includes virtual-reality and real-time 3D software development.

Yann BOUJU currently holds the position of Virtual- and Augmented-Reality Referent at DCNS. He is responsible for guidance and manages all studies concerning these technologies, for processes in industry as new systems and services for customers. His previous experience includes human-factor studies for the design of the combat system on FREMM frigates.

Lucie LEMONNIER currently holds the position of virtual- and augmented-reality project manager at CLARTE. She is responsible for leading the study and development of virtual- and augmented-reality projects. Her previous experience includes virtual-reality project development in various areas.

Chantal MAÏS is a human-factor expert. She is in charge of research in this area at DCNS Research. She is responsible for innovative exploitation with a Human-Centered Design. Her previous experience includes Human-Factor studies in virtual reality for naval ship design and Human-Centered naval ship design.

HOW MUCH FURTHER CAN SIMULATION REDUCE SEA TIME? LESSONS FROM THE ROYAL AUSTRALIAN NAVY

M Lutzhoft and **B Brooks,** University of Tasmania, Australian Maritime College, Australia.

SUMMARY

The exclusive use of traditional methods of education and training (e.g., lecture-based training programs) have failed to meet the demands of most professionalised work environments. Naval organizations across the globe have recognised this. As a result, they continue to invest heavily in the development and fielding of simulation. This investment is intended to align with the goal of reducing training and development cost and schedule while also optimizing trainee performance and learning effectiveness. We present the results of a project that aims to reduce sea time for qualification of Marine Service Technicians (MSTs) of the Royal Australian Navy (RAN) while optimizing trainee performance and learning effectiveness.

NOMENCLATURE

AMC	Australian Maritime College
CBT	Computer Based Training
MST	Marine Service Technician
RAN	Royal Australian Navy
SBT	Simulation-Based Training
VR	Virtual Reality

1. INTRODUCTION

This research was commissioned to examine several issues associated with Marine Systems Technician (MST) training. Specifically the project has considered opportunities to do more shore-based training, where efficiencies can be achieved, how these might be pedagogically appropriate and sought to identify any evidence from the peer-reviewed literature that might confirm or dis-confirm the current approach being adopted by the Royal Australian Navy (RAN) in this context.

The systems that the MSTs will be trained to operate, monitor and maintain include, the main propulsion system, auxiliary systems and the electrical generation and distribution system.

The skills that the MSTs will be expected to be proficient in, include the following; how to operate, monitor and maintain the equipment. They should be able to interpret systems drawings and flow charts and troubleshoot issues with any equipment, as well as repair any problems detected. They need a clear understanding of the process and procedures for re-commissioning equipment after maintenance and repairs have been conducted. Skills include report writing and reporting to supervisors, updating spares inventories, managing planned maintenance schedules and prioritizing work lists under a potential number of different circumstances. Furthermore, they need to know how and where to research equipment, its functions and failures, as directed.

MSTs should also know how to work with contractors, work independently, and work as a team. Finally, they should follow WHS regulations, risk assessment protocols, be capable of completing mess duties, whole ship domestic duties, security tasks and ship husbandry.

This paper mainly reports on the simulation part of this study.

2. BACKGROUND

2.1 SIMULATION AND BLENDED LEARNING

Salas, Rosen & Held [1] suggest that "In domains such as the military, medicine, business, and aviation, the present-day workplace is characterized by high degrees of complexity and competitiveness. In this environment, maximizing human performance is essential for safety, effectiveness, organizational development, and even survival" (p.328). They note that the exclusive use of traditional methods of education and training (e.g., lecture-based training programs) have failed to meet the demands of many work environments. This coincides with the increased use of simulation and learning approaches that 'blend' traditional learning, use of simulation, and other technology-supported approaches to learning. An informed decision on how much further simulation might be able to reduce sea-time can therefore be positioned in a pedagogical discussion about changes in blended learning.

"Blended learning is thus a flexible approach to course design that supports the blending of different times and places for learning, offering some of the conveniences of fully on-line courses without the complete loss of face-to face contact. The result is potentially a more robust educational experience than either traditional or fully on-line learning can offer" [2, p 64].

Military organizations across the globe have recognised this and as a result, continued to invest heavily in the development and fielding of simulation. This investment has been aligned with the goal of reducing training and development cost and schedules while also optimizing trainee performance and learning effectiveness [3]. Driscoll [4] argues that blended learning refers to four different concepts:

- To combine or mix modes of web-based technology (e.g. live virtual classroom, self-paced instruction, collaborative learning, streaming video, audio, and text) to accomplish an educational goal.
- To combine various pedagogical approaches (e.g. constructivism, behaviourism, cognitivism) to produce an optimal learning outcome with or without instructional technology.
- To combine any form of instructional technology (e.g. videotape, CD-ROM, web-based training, film) with face-to-face instructor-led training.
- To mix or combine instructional technology with actual job tasks in order to create a harmonious effect of learning and working.

However it is also clear that as military trainers begin to accept such technology, many questions remain on how to best integrate pedagogical strategies into its application [5]. The US Army has defined three classes of simulation for training: live, virtual, and constructive. Live simulations involve real people operating real systems, while virtual simulations involve real people operating simulated systems in simulated environments, and constructive simulations involve simulated people (or a representation of the behaviours of real people) operating simulated systems. As Stevens, Kincaid & Sottilare [3] note, the three classes periodically overlap each other in training, leading to the term ''blended training''.

2.2 ADVANTAGES AND DISADVANTAGES OF SIMULATION

Orlansky et al. [6] identify a detailed list of both the advantages and disadvantages of simulation, and it's easy to see from this list why so many military organisations have moved to increase their use of simulation.

Anticipated advantages include:

- Significant cost reductions achieved by not procuring and using actual equipment;
- Substitutes the use of real equipment for training;
- Allows training of high-risk tasks that would otherwise not be possible to train on;
- Provides instructional feedback to trainees;
- Reduces wear-and-tear of actual equipment, thereby extending its service life;
- Protects the environment;
- Protects operational security of tactics, techniques and procedures of using units;
- Saves time by allowing rapid scenario development, set-up, and execution;
- Saves on use of fuel, munitions, and support facilities;
- Allows novices and experts to train through the use of customizable scenarios and difficulty levels;
- May serve as a mission rehearsal tool for units through the use of geo-specific or geotypical terrain and scenarios;
- Facilitates testing and validation of plans prior to execution;
- Enables joint training of department of defense (DoD) forces and international training with coalition forces.

Disadvantages of simulation include:

- Simulation requires funding to procure, use, and sustain.
- May induce simulation sickness in trainees through the use of visual modalities and motion platforms.
- Models and simulations must accurately portray the performance of actual systems in order to avoid negative habit transfer, decreased believability, and ineffective training. This requires costly, actual data of weapons, systems, and platforms to create accurate engineering models.
- Limited fidelity can reduce training effectiveness.

From civilian experience, there are a number of benefits and drawbacks to using simulators [7]

Benefits:

- Risk Free. Studies involving risk of accidents of any kind can safely be performed. Realistic hazards in high risk operations are still present, which is important for fidelity, but associated risk is absent.
- Ethics. Studies involving performance decrements are possible in simulators, whereas it would be nearly impossible to conduct these in real life settings. Examples include alcohol, drug or medicine studies, sleep impairment, fatigue and motion sickness studies. Very stressful situations can also be explored, which would be both risky and perhaps also unethical in real world settings.
- Repeatability. A test can be repeated with or without variation almost endlessly. One should of course take into account participant dropout and/or learning effects.
- Controllability. A number of variables can be controlled and adapted to needs of the scenario. Size of vessel, place of simulation/scenario, participants, weather, sea state, visibility and other environmental aspects.
- Data assimilation. Technical data, such as video or sound, are more simply and potentially more thoroughly recorded, with enhanced control of the placement of recording devices, which may not be possible on a real ship.
- Efficiency. Control of timeline (real-time vs time-segment), e.g. travelling 200 nautical miles at 10 knots need not take 20 hours nor is there a loss of time with a ship coming up to speed.
- Cost for the use of actual vessels for research purposes can be significant. Compared to travelling to a real ship the benefits of control, less risk, and repeatability will outweigh the cost of simulation.

For some studies one may wish to validate in a ship environment, but all the groundwork can be done before, in simulators.

- Availability. Access to a specific vessel and personnel can be difficult in the shipping culture of maximising commercial gains. It may be that the research is dependent on availability of the vessel, rather than achieving the best research outcome at the appropriate time. However, availability may also be an issue for simulators. Simulators are mainly used for education purposes and likely to be well booked, and commercial projects may take priority, leaving the researcher with less favourable time slots.

Drawbacks:

- Cost of validity. Developing maritime simulated models to accurately represent their real counterpart, specific requirements or answer particular questions can be a costly and difficult process to achieve. This can require significant resources, experimentation, data acquisition and input. Multipurpose maritime simulators can be limited in their ability to meet specific requirements such as bridge layout, size, hardware and software. Producing a valid area model with photorealistic visual features is an example.
- Fidelity. Important aspects of reality may not be accurately perceived by participants, and/or a tendency to treat it like a very large and expensive computer game. However, it is often seen that after a short while, participants – especially those with more seafaring experience – will embrace the situation as if it were a real bridge watch.
- Simulator sickness. None of the studies reviewed have reported it but there is anecdotal evidence. Some instances of simulator sickness have been experienced, most likely due to the mixed signals – your eyes tell you the ship is rolling, your vestibular system tells you it is not. If the participant steps out briefly or looks at the floor, the feeling of sickness usually goes away.

2.3 MEASURING THE OUTCOMES

One of the most important questions to answer with regard to any increase in the use of simulation is: Can we achieve the learning outcomes? Beyond this, there are broader questions about the job-readiness of the trainee. Unfortunately empirical data to answer either of these questions is limited.

As Salas, Rosen & Held [1] so clearly explain, "the measurement of performance underlies the capacity of SBT [Simulation-Based Training] to prepare personnel (or evaluate their readiness) for complex environments and for organizations to manage, improve, and sustain high levels of effectiveness. Simply put, if performance is not explicitly defined and measured, it cannot be changed or improved in a systematic way" (p.329).

In defining and measuring performance, one of the key issues is the measurement of skill transfer. "The goal of a virtual simulation is generally to maximize the degree of skill transfer of the trainee from the training environment to the real or operational environment. The higher the degree of transfer, the more successful a training system is considered" [3, p. 523].

A review of 28 years of literature on surgical simulations concluded that few objective studies have been conducted that measured whether skills trained on using a surgical simulator were successfully transferred to patient care [8]. Salas, Rosen & Held [1] systematically reviewed both qualitative and quantitative methods for measurement of performance in simulation to identify best practices in this area. Their conclusion is that best practice requires more than one type of measurement, they address both the learning process and the outcomes, they describe, evaluate and diagnose performance and also identify remedial strategies to improve performance where necessary.

Part of the significant effort in assessing simulation has been driven by digging deeper into aspects of psychological performance. Often these assessments have been developed outside of normal training paradigms - such as in the testing of usability for new technologies; or the measurement of performance in operational contexts. "Strategies such as information presentation methodologies that reduce the strain on working memory while simultaneously increasing germane cognitive load, deliberate practice methodologies that aid learners in achieving fully trained status, and metacognitive prompting to improve learners' awareness of the learning process may significantly improve the utility of SBT and, consequently, trainees' abilities to apply the newly acquired knowledge. Theoretically, combinations of these various strategies may improve the effectiveness and efficiency of training" [5, p. 333].

Hontvedt & Arnseth [9] show how maritime students (cadets) interacting with a maritime pilot in a simulation take on professional roles and make joint sense of the situation. This indicates that the realism of a simulation is constructed not only by hardware and software, but also by the participants and actors in a simulation [10]. The social construction of simulated learning is therefore of particular importance.

Two studies on maritime training are of particular interest in this regard. Chauvin and colleagues [11, 12] conducted two simulator studies on situation awareness (SA) of cadets. In the 2008 study, they found few significant results relating to SA, but they saw that navigation rules and the interpretation of them were central. Over half of the participants broke the collision avoidance regulations, leading Chauvin and colleagues to devise a training program to enhance their decision making and hasten their skill acquisition in the simulator.

In the 2009 paper [12], we are presented with the somewhat surprising result that the group receiving simulator and decision making training was no better than the groups with more sea experience; in fact, the latter made safer decisions. So simulator training as a replacement for sea time may still have some way to go.

A more recent study [13] compared inexperienced and experienced boat drivers of smaller craft (e.g. power boats). Eighteen of each group drove through varying sea states (wave heights) while boat speed and eye movements were measured. In general, experts maintained a higher speed and in contrast to road driving, experts had longer fixation durations than novices. This is explained as being whereas the most dynamic aspect of a traffic scene ashore is the surroundings, at sea it is often the "driving surface".

A study by Orlandi and colleagues [14] is an example of a study with high reliability and validity. They compared plans for ship manoeuvring in port with the actual performance of those manoeuvres in a simulator. Ten marine pilots planned and subsequently performed "easy" and "difficult" manoeuvres, with variables such as cross track distance and estimated use of engine, thrusters and tugs, respectively. The results show that the methodology is valid and, for example, pilots showed a high level of adherence to plan, with variances mainly in forecasting ship position during the swing phase. Comprehensive attention to the applicability of results is evident and outcomes for both measurements of ship handling simulations and for the performance of marine pilotage are included in the results.

All of this points to the need for more research in this space – more precise measurement of the outcomes during and directly following the simulation as well as out in the work environment (job-readiness); deeper understanding of scenario design, and careful consideration of the relationship between simulation and the associated learning objectives.

3. METHOD AND PROCEDURE

This part of the report describes the implementation of a simulation learning event within the Centre for Maritime Simulations at the Australian Maritime College. By comparing the assessment of maritime tasks performed on board ship and in a simulator we took first steps toward deciding whether some task can be trained in a simulator as opposed to on board. During this pilot study we used man over board procedures using a floating object. In total 9 participants were recruited from seafaring cadet cohorts at AMC. They were 1st year students, so as not to confound findings by having previous sea experience.

3.1 BRIEFING AND TRAINING

All students received a joint task briefing for about 30 minutes. Following that they split into three groups for task training sessions led by different instructors, for 2-3 hours, at one of the following locations:

- Reviresco (14 meter school training vessel), 3 students
- Full mission simulator AMC, 3 students
- ECDIS laboratory AMC "Computer based training lab", 3 students

The students planned and performed a manoeuvre designed to get as close to the object as possible. On the ship there was a normal crew (a master and a deckhand). In the full mission and CBT groups, the same task applied (but no crew).

3.2 ASSESSMENT

All 9 students participated in a final session on the Reviresco for task assessment, lasting about 20 minutes per person. Two observers were present to evaluate the plan and manoeuvre on the ship, using a set of assessment criteria from normal AMC examination methods. The criteria used were Wind speed and direction assessed, Tide and current, Position, General manoeuvring, Approach to MOB, Time for manoeuvre, Team Work and Preparations, see example in figure 1.

Criterion	5	4	3	2	1	0
Wind speed and direction assessed	Direction and type of turn assessed taking into account the wind speed and direction	Wind speed and direction noted at start of turn – type of turn not appropriate for the wind direction and strength	Wind strength and direction assessed and the approach adjusted after the initial turn	Wind direction noted, and final approach adjusted accordingly	Wind direction only noted in the final stages of the approach when attempting to pick up the MOB	No account taken of the wind direction and effect

Figure 1: Example of assessment criteria

4. RESULTS

The results of the assessment were grouped according to the criteria.

Group 1 – trained and assessed on Reviresco, generally performed excellently, at the high end of the criteria. The only lack was in preparation.
Group 2 – trained in full mission simulator, assessed on Reviresco, performed very well, some grades a little lower on teamwork and preparation.
Group 3 – trained in CBT lab, assessed on Reviresco, had more scattered results, starting at mid-grade and moving towards the low end on all rubrics. Two persons would have failed, had it been a real assessment.

5. DISCUSSION

Blended learning is a flexible approach to course design that supports the blending of different times and places for learning, offering some of the conveniences of fully on-line courses without the complete loss of face-to face contact. In the context of the current project, the alteration of the 'blend' includes the potential for altering the balance between shore-based and ship-based training, noting that this is also likely to bring with it an increase in the use of technologies that provide virtual access to the ship such as simulation and virtual reality.

This is not, however, a guaranteed solution to improve learning outcomes. "While noting its considerable potential to improve the student learning experience, we have found that blended learning is not automatically successful" [2, p.78]. These authors note that the integration between online and face-to-face elements of the learning environment, cultural appropriateness and administrative support are key factors in success, but also indicate that their experience suggests blended learning does not necessarily lead to a reduction in face-to-face education.

A review of the effectiveness of e-learning strategies in the Australian Army, Newton & Ellis [15] placed this learning challenge in a broader context. They identify the need to address several factors including, but not limited to the needs of the learners. These include alignment with organisational goals, dynamic response to change in the organisation. It is apparent that a strong 'build it and they will learn' approach exists with respect to the use of simulation.

Fidelity or realism may also be costly and it should be carefully considered if the added benefit is a return on investment. The demand for high realism seems to come mainly from experienced seafarers, whereas simpler training of routine type or specific tasks may not need a high degree of fidelity. However, for familiarisation and active operational tasks, higher realism is needed, and immersion aspects also become more important.

It is clear that strong reasons exist to engage a blended learning model and that these reasons out-number the reasons to apply only traditional models. Indeed the vast majority of industries and occupations discovered this a long time ago, and are simply adjusting their blend as the context dictates they should.

In order to establish what works and what does not it will be necessary to measure effectiveness, and to measure it in multiple ways. Kirkpatrick (cited in [16]) offers four levels at which we should assess training effectiveness and at some stage when the 'blend' has been bedded down and is relatively stable, an assessment along these lines is warranted.

The principles of good educational and training design does not change, however. The designer still needs to acknowledge the fundamentals of learning objectives, visible learning, linking assessment to objectives and addressing the need for authenticity in assessment and transferability of skills to be as 'job-ready' as possible. This means that also the trainers need to be 'up-skilled'. Furthermore, some aspects of a blended approach that is heavier in the use of technology might see the educator/trainer immersed in multi-disciplinary design teams, and playing the role of 'coach' to provide a bridge between the learning environment and the workplace.

Finally, new technologies are being implemented to support blended learning approaches. According to the available literature it is perhaps too early in their development and use to truly understand their value. As they become more embedded that value will become more obvious, and could reasonably be linked with the assessment identified above. There is little peer-reviewed literature that demonstrates an empirical relationship between the increased use of educational technologies and improved graduate job-readiness. There is also little peer-reviewed evidence to support:

- Increases or decreases in time spent on training using simulation.
- Improvements or decrements in training quality and by associated job-readiness associated with increases in shore-based (or on-the-job) training.

This is in part because the technology is relatively new, so the studies have not been performed, however it is also true that educational research sometimes lacks the reliability and validity to be able to demonstrate such relationships. Still, this situation is complicated by the need to be able to track individuals from trainee to graduate to workplace performance.

As the relationship becomes more specific – for example assessing particular cohorts (maritime engineers, or naval personnel) for changes in job readiness associated with altering blends of learning technologies – the literature is incredibly scarce and points to the need to progress the RAN approach carefully and systematically, creating a relevance evidence-base as RAN proceed.

There are at least two situations where time saving may be possible.

1. Compression of shore-based training is likely to be achieved by using technology such as high fidelity simulation and VR. Anecdotal evidence from the cruise industry indicates that time spent in simulation as preparation for entering new situations (e.g. a new port) may halve the total training time, with the obvious benefit of "recognising" the situation at first real encounter. This may gainfully be employed in systems training, team training, familiarisation and induction of MSTs.

2. Shorter time spent on board to consolidate and validate the on shore training. This time is estimated to be able to reduce training allocation to 50% of earlier needs, given good pedagogic design of the training (in combination with, and as validation of, the above mentioned technology-supported shore training).

The above will also address the long waiting time for getting "sea-time" and only a short on board period is needed to become job ready. Reducing the period of "waiting for sea time" between the completion of training and commencing their position on board a ship for the endorsement of their qualifications may be addressed by a new training regime. If familiarisation and similar training can be performed with the VR walkthroughs the MST may be occupied with getting to know the ship.

6. CONCLUSIONS

As organisations alter the way they train, a review of instructor training also becomes important. Such a review would examine the length, methods and the effectiveness of the teaching techniques, including modern pedagogy, that are taught to instructors. Additional and up-skilled instructors will be required to deliver the increased simulator training.

Reducing the period of "waiting for sea time" between the completion of training and commencing their MST position on board a ship for the endorsement of their qualifications may also be addressed by a new training regime. If familiarisation and similar training can be performed with VR walkthroughs the MST is occupied and spending time getting to know the ship.

Non-technical skills related to teamwork and collaboration within and outside of ship need to be included as part of the training – for instance working as part of an emergency team, providing leadership and peer support or working with contractors. Utilisation of proper communication channels, and knowing what to communicate, requires practice in every learning environment including the classroom, workshop, and in simulation training, is necessary to ensure that the practice continues on board the ships.

6.1 SIMULATION

It is reasonable to assume that some "practical tasks" can be trained in full mission simulators or high fidelity environments. It is also clear that more research is needed on which types of tasks can be trained. The Chauvin studies [11, 12] seems to indicate the results are similar for decision making skills trained in a simulator or at sea. What we cannot say at the present time is whether the final assessment (or part of it) can be made ashore as well.

Our last comment reflects the somewhat pragmatic and organic growth in the use of simulation. "My experience is that folks will buy, for example, something like a high-fidelity simulator through a grant," says Chris Boyer, MPA, NRP, FP-C, simulation coordinator at Delaware Technical Community College in Dover. "But when they buy the simulator, they never actually incorporate a cost for somebody to come in and train people how to use it. So they're trying to learn how to use it on the fly, on top of trying to write the scenarios and run the simulations, and they never get comfortable with it." [17, p. 18] The key is to make sure simulation is used and learning scenarios built around clear, succinct learning objectives [17].

7. ACKNOWLEDGEMENTS

We acknowledge the cooperation of the RAN, and the assistance of Julie Binder, Gamini Lokuketagoda, Mohan Anantharaman, Paul Brown, Clarence Pietersz and Siriwardhana Gamage.

The project was supported by grant 00003994. The simulation part of this project was performed with ethical clearance from Tasmania Social Sciences Human Research Ethics Committee, no H0015632.

The views presented in this report are the authors' and do not necessarily reflect the views of the RAN.

8. REFERENCES

1. Salas, E., Rosen, M., & Held, J. Performance measurement in simulation-based training: a review and best practices. *Simulat Gaming* (2009) 40: 328–376.

2. O'Connor, C., Mortimer, D. & Bond, S. Blended learning: Issues, Benefits and Challenges. *International Journal of Employment Studies* (2011) 19(2): 62-82.

3. Stevens, J., Kincaid, P. & Sottilare, R. Visual Modality Research in Virtual and Mixed Reality Simulation. *Journal of Defense Modeling & Simulation* (2015) 12(4): 519-537.

4. Driscoll, M. Blended learning: let's get beyond the hype. (2002, March 1). Retrieved Aug 30, 2010, Elearning: http://elearningmag.com/ltimagazine.

5. Vogel-Walcutt, J., Carper, T., Bowers, C. & Nicholson, D. Increasing Efficiency in Military Learning: Theoretical Considerations and Practical Applications. *Military Psychology* (2010) Vol. 22, No. 3: 311-339.

6. Orlansky J, Dahlman C, & Hammon C. The value of simulation for training. Institute for Defense Analyses, Alexandria (1994).

7. Lutzhoft, M., Brown, P., Dunham, R., & van Leeuwen, W. Maritime research. In M. Young (ed) Simulators for Transportation Human Factors: Research and Practice. Farnham: Ashgate (In press).

8. Schout, B., Hendrikx, A., & Scheele F. Validation and implementation of surgical simulators: a critical review of present, past, and future. Surgical Endoscopy (2010) 24: 536–546.

9. Hontvedt, M., & Arnseth, H. C. On the bridge to learn: Analysing the social organization of nautical instruction in a ship simulator. *Computer-Supported Collaborative Learning* (2013) 8, 89-112.

10. Lützhöft, M., Porathe, T., Jenvald, J., & Dahlman, J. System Simulations for Safety. In O. Turan, J. Bos, J. Stark, & J. L. Colwell (Eds.), *Human Performance at Sea*. Glasgow, Scotland, UK: University of Strathclyde (2010) 3-10.

11. Chauvin, C., Clostermann, J. P., & Hoc, J.-M. Situation Awareness and the Decision-Making Process in a Dynamic Situation: Avoiding Collisions at Sea. *Journal of Cognitive Engineering and Decision Making* (2008) *2*, 1-23.

12. Chauvin, C., Clostermann, J. P., & Hoc, J. M. Impact of training programs on decision-making and situation awareness of trainee watch officers. *Safety Science* (2009) 47(9), 1222-1231.

13. Godwin, H. J., Hyde, S., Taunton, D., Calver, J., Blake, J. I. R., & Liversedge, S. P. The Influence of Expertise on Maritime Driving Behaviour. *Applied Cognitive Psychology* (2013) *27*, 483-492

14. Orlandi, L., Brooks, B., & Bowles, M. A Comparison of Marine Pilots' Planning and Manoeuvring Skills: Uncovering Mental Models to Assess Shiphandling and Explore Expertise. *The Journal of Navigation* (2015) 68(05), 897-914.

15. Newton, D. & Ellis, A. Effective Implementation of E-Learning: A case study of the Australian Army. *Journal of Workplace Learning* (2005) 17(5), 13.

16. Shobe, K. & Curtis, B. Training assessment challenges: In a military environment a U.S. Navy Example. Performance Improvement (2007) 42-46

17. Erich, J. The Secrets of a Successful Simulation Program. *EMS World* (2016).

9. AUTHORS BIOGRAPHY

Margareta Lützhöft is a master mariner, trained at Kalmar Maritime Academy in Sweden. After leaving the sea, she studied for a Bachelor's degree in Cognitive science and a Master's in Computer Science. In 2004 she received a PhD in Human-Machine Interaction. Presently she is holding a position as Professor of Nautical studies at the Australian Maritime College. Her research interests include human-centred design and the effects of new technology.

Benjamin Brooks is a maritime and emergency management human factors researcher. He has a professional background in Hazardous and Noxious Substance (HNS) consulting and was a safety manager for a large healthcare provider before joining academia. He has a PhD in Psychology. His current role is Research Fellow at the Australian Maritime College. His research interests are in all things safety management and human performance in extreme environments.

THE ROLE OF NAVIGATIONAL ASSISTANCE IN MANAGING UNCERTAINTY AND IMPROVING SAFETY AT SEA

L de Vries, Chalmers University of Technology, Department of Shipping and Marine Technology, Sweden

SUMMARY

This paper considers a case study of a potential close quarters situation between three vessels, viewed as a case of sociotechnical work and analysed using a model of navigational assistance developed using the Functional Resonance Analysis Method (FRAM). A large bulk carrier finds itself in a potential close quarters situation with two other vessels near the entrance to a deep water channel. Uncertainty regarding the vessel's draught restricts its manoeuvrability, increasing the risk of collision and grounding. Analysis of this case illustrates how a potentially unsafe situation may be identified, communicated and resolved by the interaction between vessels and shore, thereby improving safety in everyday operations. Furthermore, it shows how FRAM may be used effectively both for modelling a work practice on a generic level, and also for understanding how work is conducted safely in a specific situation.

NOMENCLATURE

[Acronym/Unit]	[Definition]
AIS	Automatic Identification System
FRAM	Functional Resonance Analysis Method
HF	Human Factors
IALA	International Association of Marine Aids to Navigation and Lighthouse Authorities
IMO	International Maritime Organisation
IMPA	International Maritime Pilots Association
m	Metres
nm	Nautical miles
RE	Resilience Engineering
STS	Science and Technology Studies
VHF	Very High Frequency (radio)
VTS	Vessel Traffic Services
VTSO	Vessel Traffic Services operator
WS	Workplace Studies

1. INTRODUCTION

"It is by God's Almighty Providence and great chance, and the wideness of the sea, that there are not a great many more misfortunes and ill chances in navigation than there are."

- Samuel Pepys, 1683

Navigation of seagoing vessels is one of the oldest, and most hazardous, professions on earth; the use of local expertise in safely navigating dangerous waters has been documented since Greek and Roman times [1]. Traditionally these services have been provided on board the vessel by maritime pilots [1, 2], and more recently from shore by Vessel Traffic Services (VTS) operators [3, 4]. Continuous advancements in navigation, positioning and communication technology have created new possibilities to provide additional assistance both on board and from shore. The maritime domain is increasingly viewed as a single sociotechnical work system [5], and ship-shore interaction is being developed under such names as e-navigation [6], sea traffic management [7], and maritime traffic management [8]. The International Maritime Organisation (IMO), whose primary purpose is to promote "safe, secure and efficient shipping", aims to harness the potential of such developments to "facilitate a holistic approach to the interaction between shipboard and shore-based users, under an overarching e-navigation architecture" by 2019 [6].

According to the IMO, pilotage aims "to guide vessels into or out of port safely - or wherever navigation may be considered hazardous, particularly when a shipmaster is unfamiliar with the area" [2]; likewise the goal of VTS is to "improve the safety and efficiency of vessel traffic and to protect the environment" [4]. This paper explores the contribution of pilots and VTS operators towards safety at sea, under the collective term *navigational assistance*. Using a case study of a typical navigational situation, it aims to understand how navigational assistance is performed in everyday operations, and how pilots and VTS operators deal with uncertainty and variability in their work to ensure a safe outcome.

This will be done with a sociotechnical systems approach using the Functional Resonance Analysis Method (FRAM) [9, 10], and informed by the traditions of Science and Technology Studies (STS) [11] and Workplace Studies (WS) [12, 13]. If successful, this approach may provide a way to capture operator knowledge which may be used to inform the design of future e-navigation systems.

2. BACKGROUND

Within the fields of Science and Technology Studies (STS) and Workplace Studies (WS) there is a rich tradition of observing how sociotechnical work is conducted in everyday situations. Predominantly qualitative methods such as observations and interviews are used to "make work visible" [13] by describing in detail the network of interactions between humans and technology, usually distributed in time and space, to deal with emerging, often uncertain, circumstances. Examples of this type of approach can be found in safety-critical domains such as aircraft ground control [12], aviation [14], navigation [15] and offshore [16], and in general safety research [17]. A central theme of STS/WS is that the configuration of sociotechnical work is not fixed; networks emerge from the interaction of human and non-human actors (e.g. technology), which are themselves transformed by the process of interaction [11, 18]. Since non-humans are actors, they may also be objects for observation in empirical studies [11, 18]. Uncertainty is seen as an inherent property of the actors, their interactions *and* the resulting outcome; it is a consequence of the open nature of networks, allowing actors to adapt to the circumstances [11, 19].

Within the Organisational Safety field, work, interaction, safety and uncertainty are more commonly viewed through the lens of sociotechnical systems [20-24]. Systems are characterised by non-linear interaction between components, with a focus on variability and control [21, 22]. Many writers have expounded on the apparent tension between safety in principle (i.e. the "official" version described in procedures, safety management systems, risk assessments etc.) and safety in practice (i.e. resulting from the ability of humans and organisations to adapt to circumstances), summarised in [23]. Recent developments such as Resilience Engineering [22] and Safety-II [24] seek to bridge this gap by examining "work as done", rather than "work as imagined" to understand how variability in a work system may be managed to ensure successful outcomes [24].

Despite obvious parallels, STS/WS and systems theories are usually treated quite separately in literature [19]. Organisational Safety is linked implicitly with the latter - e.g. in the work of Rasmussen and Hollnagel - rather than the former, with few exceptions [e.g. 16, 17, 25]. Uncertainty and variability are prominent features of both, but are applied very differently. STS/WS does not provide a clear definition of uncertainty, but describes it in terms of interaction and formation of networks [11, 12]. Variability - as commonly used in Resilience Engineering and Safety-II - is, according to probability theory [26], only one of many forms of uncertainty (aleatoric uncertainty). Lack of knowledge (epistemic uncertainty) is another common form which is often present in sociotechnical work.

2.1 THE FUNCTIONAL RESONANCE ANALYSIS METHOD (FRAM)

The Functional Resonance Analysis Method (FRAM) [9, 10] is described as "an analysis tool that reflects Resilience Engineering and Safety-II thinking" [10: p12]. It models a sociotechnical system in terms of its functions, and the interactions between these, in order to analyse where variability may arise and "resonate" throughout the system, leading to potentially unwanted outcomes. Sources of variability may be either endogenous (internal), exogenous (external) or upstream/downstream (one function affects another); effects of variability may be expressed in terms of time or precision, or failure modes [9: pp63-75].

The FRAM method [9, 10] may be summarised as:

- Step 0: Recognise the purpose of the analysis i.e. to understand the practice of navigational assistance and describe this using a FRAM model.
- Step 1: Identify and describe the functions. A function is the "means that are necessary to achieve a goal" or the "activities - or set of activities - which are required to produce a certain outcome" [9: pp40-41].
- Step 2: Identification of variability. The functions should be examined separately to determine how each of them may vary, due to either endogenous, exogenous or upstream-downstream variability.
- Step 3: Aggregation of variability. Potential couplings between functions should be analysed to discover how variability in one function may affect others, and the outcome.
- Step 4: Consequences of the analysis. Potential ways to manage or control variability discovered during the previous steps should be analysed.

Hollnagel [9: p33] originally envisaged FRAM as a two-stage process: "The first is using the FRAM to develop a model of the activity (process or performance) that is the focus of the analysis [steps 0-1]. The second is to use the model to create instantiations of the activity (or performance) and then to analyse these [steps 2-4]." FRAM has mainly been used for analysis of single events or accidents [e.g. 27, 28]; here both model and instantiations were created with a particular event in mind. It has also been used to investigate work systems - such as VTS [29] and oil response [30] - in terms of their capability for resilience, performing the analysis of variability at the system level, without reference to a specific event or situation.

This paper aims to go one stage further by using a model previously developed at the system level [31] to investigate an individual case (or instantiation), thus completing both stages of Hollnagel's process at the macro- and micro-level. It will also borrow some ideas from STS/WS which the author believes will bring added value to a FRAM analysis. It thereby aims to use the "different ways of seeing" [25] associated with the respective traditions to improve our ability to describe a practice such as navigational assistance. These are: following an object (a vessel) [11, 18], and; the formation of a network (pilot-vessels-VTS) [11, 12]. It will also employ a broader concept of uncertainty (including both variability and uncertainty due to lack of knowledge) [26].

3. SETTING THE SCENE: PREVIOUS STUDIES ON NAVIGATIONAL ASSISTANCE

The case studied here forms part of a larger body of ongoing work on navigational assistance, which so far encompasses empirical studies of pilotage and VTS in eight pilotage/VTS areas in four European countries [31, 32]. Data collection has used qualitative methods such as focus groups, individual and group interviews and field observations, derived from [18, 33]. Data were analysed using a grounded theory-inspired thematic analysis approach based on the principles of iteration and emergence [34], open coding [35] and axial coding [36].

The codified data from this empirical analysis were subsequently used to construct a generic model of the practice of navigational assistance using the FRAM method as described above, and following the guidelines provided in [9, 10]. All steps (0-4) were performed on the system i.e. generic level of navigational assistance [31], similar to [29, 30]. This produced a model of the common elements found in the work of pilots and VTS operators, and regarded by them as essential for successful assistance, whether provided on board or from ashore.

In summary, pilots and VTS operators described their work as being dependent on: (i) communication and trust between the pilot, VTS operator, and the master and crew of the vessel, and; (ii) the use of local knowledge, preparation and foresight to integrate information from a wide range of sources and provide timely assistance to vessels. Local knowledge, preparation and foresight can each be sub-divided into factors relating to vessels and traffic, the environment, and human and organisational issues. Successful assistance was defined as "no incidents", but also actions to avoid or mitigate incidents, even when these resulted in a negative outcome. "Incidents" ranged from actual accidents, such as collision or grounding, to potential accidents, such as approaching shallow water or exceeding (draught or air draught) limits. Previous work on pilotage [37], VTS [29] and ship-shore interaction [8], has highlighted many of the same features as integral to navigational assistance.

The FRAM model (Fig. 1a, full page version, Fig. 1b) [see also 31] shows how local knowledge (top left), preparation (middle left), foresight (bottom left) are used to integrate clusters of vessel/traffic, environmental and human/organisational functions, which are communicated between vessels and shore by the pilot (top right, red) and VTS (bottom right, green). These functions are listed in Table 1. All functions may be performed by either pilot and/or VTS operator, except four 'pilot' functions, which are usually performed by the pilot only (outlined in purple in the model).

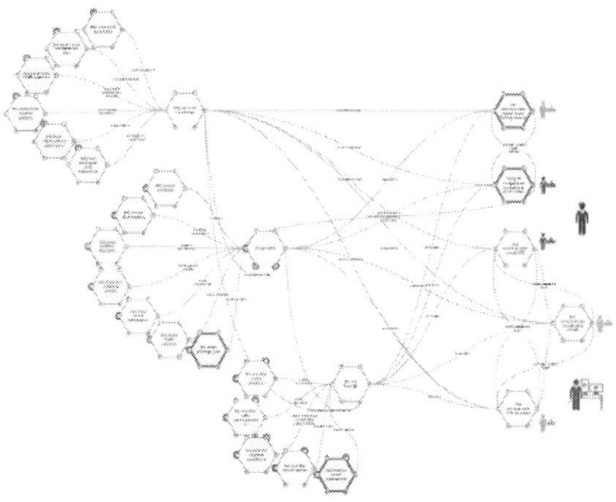

Figure 1a: Generic FRAM model of navigational assistance.

Table 1: List of FRAM functions.

Function name
(to) use local knowledge
cluster functions:
(to) know local geography
(to) know local navigational aids
(to) know local traffic patterns
(to) know local weather patterns
(to) have shiphandling experience
(to) have pilotage/VTS experience
(to) prepare
(to) receive handover
(to) receive pilot booking
(to) check weather forecasts
(to) check in/outgoing vessels
(to) check vessel information
(to) check traffic situation
(to) make pilotage plan (pilot function)
(to) use foresight
(to) monitor traffic situation
(to) monitor radio communication
(to) monitor weather conditions
(to) monitor vessel motion
(to) monitor vessel instruments (pilot function)
(to) communicate vessel-VTS
(to) communicate VTS vessel(s)
(to) communicate vessel-other vessels
(to) give navigational instructions pilot-vessel (pilot function)
(to) communicate vessel-tugs/fishing vessels (pilot function)

How navigational assistance is actually performed, which actions are deemed appropriate and ultimately what is considered "safe" were described as being to a large degree situation-dependent. The combination of vessel/traffic, environmental and human/organisational factors varies in each situation, as do many of the individual factors, both due to inherent variability but also how they are known and understood by the actors. The ability to foresee how a particular situation may develop and adapt to this is a critical element of successful assistance. Accurate pre-information about vessels is therefore essential, but was often seen as a cause for concern by both pilots and VTS operators.

3.1 DEPTH AND DRAUGHT: A COMMON CONCERN FOR THE SAFETY OF NAVIGATION

To illustrate the practice of navigational assistance, we will investigate the importance of accurate depth and draught information, which was a recurring theme throughout all the empirical studies, not just the case in question. The information in this section is a synthesis of information gained from pilots and VTS operators during the author's previous studies; this will be explored in detail using the case study and instantiations of the FRAM model.

In all areas, narrow channels and shallow waters, combined with variations in water depth due to tide or water level, mean that vessels with larger draughts may be restricted in their ability to navigate safely. Although actual groundings were not seen as common, "heading towards shallow water" was viewed as a regular occurrence. It was also seen as a typical situation where proactive assistance on the part of pilots but particularly VTS operators was seen as producing a noticeable improvement to the safety of navigation. In one area, VTS operators reported that, *"last year we had about 20-25 vessels heading for shallow water, and it's a potential grounding"* but *"if a vessel is heading towards shallow water and the VTS calls them and says so, that's very good. Then they have time to react so they don't go aground"*.

In one area, semi-diurnal tidal variations and strong currents restrict entry to the port through the breakwaters. Deep draught or channel-bound vessels are dependent on a tidal window to pass the breakwaters. The tidal advice for each specific vessel is produced by software in the VTS with input from forecasts obtained from the national meteorological institute, and communicated to the vessel by the pilot. The estimated time of arrival (ETA) at the breakwaters must be accurate, and correspond with the recommended time of arrival (RTA) indicated in the tidal advice, since there is a "point of no return", beyond which there is no space to turn and vessels must continue regardless of available water depth.

In another area, nearby pressure systems and prolonged periods of high wind may cause water to be transported away from the area, reducing the water depth in narrow channels passing through the area. Vessels over a certain draught, which usually transit through the area or enter into ports, may find they either have to anchor and wait for the sea level to return to normal, or to take another route. Both of these alternatives may cause delays and additional expense on the part of the ship or cargo owner. Although this is not a frequent occurrence, it happens several times a year, and so the pilots and VTS must be prepared for it.

Accurate information about the vessel draught (i.e. how far the hull extends below the waterline) is therefore essential in planning which routes vessels may take and at which times they may enter and leave. Each vessel's draught may however vary with the loading condition and change over time as fuel is consumed. For an average bulk carrier of about 180 metres length (around 50,000 deadweight tonnes) for example, this may vary between about five metres and thirteen. This figure may however be communicated in various ways: it may arrive via the pilot planner, by web form, email or telephone call from the agent or shipowner, or from the vessel itself; it may be sent directly to the VTS by the same means. The source of this measurement will usually be unknown to the receiver. The vessel may also transmit its draught via AIS (Automatic Identification System) and should announce it over the VHF radio before pilot boarding and when entering the VTS area.

While the vessel has an actual draught at any given time, measured on board by sensors (on newer vessels) or visible on load lines painted on the side of the vessel, this updated value may not necessarily be the one which is communicated in advance or transmitted by the vessel's AIS transponder. Once on board, the pilot may check the load lines or instruments and discover the vessel has another draught than reported. Despite being essential for determining where a vessel may or may not go, inaccurate or conflicting draught information is not uncommon and is seen as a source of uncertainty by the majority of pilots and VTS operators.

4. METHODOLOGY

4.1 CASE STUDY: EMPIRICAL OBSERVATIONS OF VTS OPERATORS AND PILOT AT WORK

The case used here concerns a potential close quarters situation between three vessels, which was witnessed by the author while conducting field observations and workplace interviews at a VTS centre. The direct participants were three VTS operators and one VTS manager; a pilot and several vessels were indirectly involved. Data collected were: voice recordings and transcripts of conversations with on-watch VTS operators; VHF radio traffic between the VTS and vessels; pictures of the VTS combined chart and radar displays; vessel information in the VTS electronic ship report system, in emails from the vessels, written notes taken by the VTS operators, and online. The observation strategy may be described as "following the vessel",

derived from the concept of shadowing objects found within STS [11, 18].
Textual, visual and audio material from the case, in the form of sound files, transcriptions, drawings, photographs and written notes, were analysed using the same thematic analysis technique described above (Section 3), and a synthesis will be presented in narrative form below. This also formed the basis for the FRAM analysis in Section 6.

4.2 FRAM ANALYSIS: IDENTIFICATION, PROPAGATION AND MANAGEMENT OF VARIABILITY

The FRAM model presented in Section 3 [31] was used to analyse the case in order to understand how variability manifested itself and spread through the ship-shore system in this particular situation. This analysis will be presented in detail in Section 6 below. The analysis involved retaining the model functions developed in step 1 of the FRAM method (above), and repeating steps 2-4 (identify, aggregate and manage variability) using the empirical material from the case study; essentially mapping the case onto the generic FRAM model to produce instantiations of the model. This was done by once again "following the vessel" [11, 18], tracing how networks of interactions formed [11, 12, 18] as the situation developed, but focusing on the interactions between the functions, rather than the actors.

Though, as Checkland [38] noted, the study of sociotechnical systems is often too "messy" to be neatly described. This was found when attempting to analyse the whole case using this technique. To present the propagation of variability, we will therefore "zoom in" to certain key functions to create instantiations showing interactions at three important stages of the case study.

5. CASE STUDY: AN EVERYDAY NAVIGATIONAL SITUATION

The case we will investigate takes place in an area in which several alternative north- and south-going routes may be taken by vessels, with the actual route chosen depending largely on the draught of the vessel. A nautical chart (Fig. 2) shows entrance to the southbound deep water route (marked with red circle), first passing the westerly cardinal mark, then going between the red buoys to the east and the easterly cardinal mark to the west, before continuing to the south-east. On either side of this is shallow water, shown as dark blue (less than 10m depth) and light blue (less than 20m, mainly around 12m) on the nautical chart. This information is not immediately obvious to the VTS operator from their electronic chart display (see Fig. 4 below), or indeed at sea. However in both cases the buoys and navigational marks are visible; pilots, VTSOs and ships which regularly traffic the area know their significance.

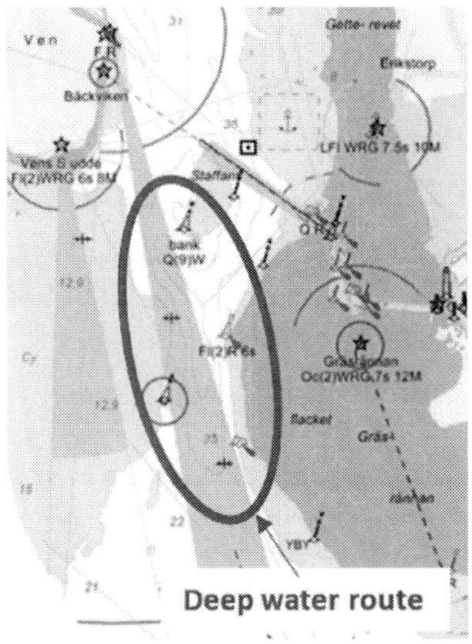

Figure 2: Nautical chart showing area of interest.

This area is considered by pilots and VTS operators working there to be one of their "hot spots", where potentially unsafe situations regularly occur. For example:

"Last week we had an incident. They're [vessels] coming from here [south-west] going north. Sometimes they will take this way around the buoy and go up here. And then you have the situation with the southbound vessels that are going down here [east of the island]. Last week we had two vessels coming southbound. One had a pilot on board, deep draught, couldn't turn to starboard. And we had one also coming up. He wouldn't turn. The pilot called and called and called, 'Please turn. I cannot turn. I'm deep draught.' He was calling for five or ten minutes trying to get the other vessel to turn. Then he called us and he wanted to report it, and we just told the pilot 'Yes, we're recording. We will make a report and we will also make contact with the ship'. And we did it and eventually he turned and there was a CPA [closest point of approach] of 0.2 [nautical miles]. It was a little bit close."

Our case also involves three vessels in a similar situation, with two southbound, one of which is deep draught, and a northbound vessel (Fig. 3) (names given are pseudonyms):

- ALPHA, a general cargo vessel with draught 5.1m, heading south through the area.
- BETA, an oil products tanker with draught 5.5m, which regularly transits between two ports close to the area. BETA holds a Pilot Exemption Certificate, meaning that, due to competence of the crew and familiarity with the area, there is no requirement to have a pilot on board.

- CHARLIE, a large bulk carrier, passing through the area, heading for a port to the south-east.

Figure 3: Vessels of type ALPHA (top), BETA (middle) and CHARLIE (bottom), not to scale, source: MarineTraffic.com.

A rule of navigation will also come into play, namely COLREGS [39] Rule 14, which states that when two vessels are in a head-on situation with risk of collision "each shall alter her course to starboard [right] so that each shall pass on the port [left] side of the other".

CHARLIE has taken a pilot on board prior to entering the VTS area from the north. They have previously sent an email to the VTS with the vessel draught noted as 7.4m. The VTS operator has checked the MarineTraffic website and noticed that the registered draught of the vessel is 12.8m. The pilot makes radio contact with the VTS.

***VTSO**: Just to confirm your draft, 7.4?*
***CHARLIE (PILOT)**: No, draft is now 7.15.*
***VTSO**: 7.15. OK, that's well copied.*
***CHARLIE**: Yes, there has been a de-ballasting issue. They thought the draught was OK.*
***VTSO**: OK, can you get one of the mates to update the AIS transponder now?*
***CHARLIE**: Yes, I will try.*
***VTSO**: Thanks for that.*

On passing the reporting line for the VTS area, the pilot then makes the ship report via VHF radio. Draught is now shown on the VTS screen is 7.1m. This information is AIS (Automatic Identification System) data, input by the crew and transmitted via satellite to the VTS and other vessels. Being uncertain whether or not 7.1m is CHARLIE's true draught, the pilot informs the VTS that they will take the deep water route on the eastern side of the area. (The draught limit for the channel on the next stage of the journey is 7.2m.)

***CHARLIE**: [VTS], [VTS], confirm.*
***VTSO**: This is [VTS], good morning.*
***CHARLIE**: Good morning [VTS]. Pilot on board CHARLIE, crossing [buoy] southbound, passing east of [island] and [] Channel.*
***VTSO**: Pilot on board, east of [island] and [] Channel. Okay, thank you, good watch.*

***VTSO** to author: Now we have the [] pilot on board.*

Approaching the entrance to the deep water route, it becomes apparent that CHARLIE will find itself in a potential close quarters situation with ALPHA and BETA. First it faces a tight passage between ALPHA to the west and shallow water east of the westerly cardinal mark. Then it will meet BETA in a head-on situation, but altering course to starboard (in accordance with COLREGS Rule 14) will put it close to shallow water to the west of the intended route.

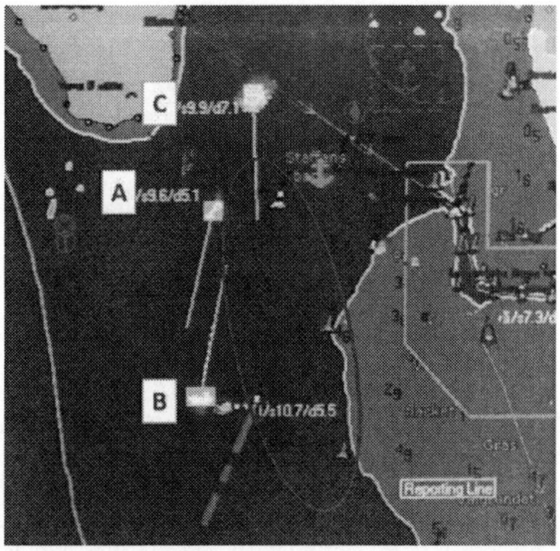

Figure 4: VTS display showing vessels and entrance to deep water route.

At this point the VTS operator is monitoring the screen (Fig. 4). (Note that depth on the VTS screen is show as light blue (less than 8m) and dark blue (over 8m).) The author wonders why he does not call up the vessels on the VHF and request their intentions. He explains that, since *"[CHARLIE] has a [local] pilot on board... [BETA] is [local] also"*, he is confident that they will sort out the situation themselves. First, CHARLIE calls ALPHA on the VHF.

***CHARLIE**: ALPHA... CHARLIE...*
***CHARLIE**: Hello ALPHA Sir 7-3.*
***ALPHA**: Yes. I'm on ALPHA.*
***CHARLIE**: ... Good morning captain. This is the bulk carrier CHARLIE astern of you. We are overtaking you and my intention is to overtake you on the port side ... I will come a little bit [...] west cardinal buoy.*
***ALPHA**: OK, you will overtake me from my port side. No problem.*
***CHARLIE**: You can please give me some more space. Take some more degrees to starboard, OK?*
***ALPHA**: Of course. I will do.*

When asked if it is normal that the traffic organize themselves, the VTS operator replies, *"Just let them talk. It's better. ... I know that, if the vessels, on the bridge, are not worried, why should I sit here and be worried? ... But if there is something odd then I call them up. ... It's nice weather. They should be able to see each other."*

BETA then calls CHARLIE on the radio:

***BETA**: CHARLIE - BETA 7-3.*
***CHARLIE**: BETA - CHARLIE*
***BETA**: Hello. Shall we meet... This is BETA. Shall we meet starboard-to-starboard?*
***CHARLIE**: Yes, that is very good. We'll also be a bit more to port when we're coming so that will be very nice [?]*
***BETA**: OK. Well received. Starboard-to-starboard.*

The VTS operator listens to this exchange and says that *"they have been talking so they know what is going on, they will meet starboard to starboard"*. He continues to monitor until CHARLIE has overtaken ALPHA and passed BETA safely.

6. FRAM ANALYSIS: INSTANTIATIONS OF EVERYDAY WORK

Since the purpose of the FRAM analysis is to follow how variability resonates throughout the ship-shore system, we will start with the parameter which is of primary concern in this situation - the vessel draught. Having followed vessel CHARLIE in the case study narrative, we will do the same in the FRAM analysis. Three FRAM instantiations will be created, showing how variability or uncertainty is identified, propagated and managed at three key stages of the case study. Instantiations are shown in this section without annotated connections between functions; variability is shown by a sine wave on the affected function. Please also see detailed illustrations at the end of the paper (Fig. 5b, 6b, 7b).

6.1 IDENTIFICATION OF VARIABILITY

Uncertainty regarding CHARLIE's draught is discovered by both the pilot and the VTS operator separately when each is performing the function *'(to) check vessel information'*. Table 2 shows how this function *should* operate, however as noted above, it is completely normal that pre-information is either not received, incorrect or conflicting, and thus the output is variable and the vessel information, even after checking, is still uncertain.

Table 2: FRAM function *'(to) check vessel information'*.

Function	(to) check vessel information
Description 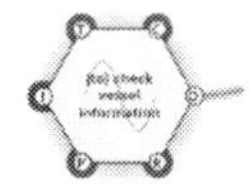	Pilot/VTSO checks vessel information to ensure correct information received and vessel fulfils requirements for safe passage on intended route.
Aspect	
Input	vessel pre-information received from: - pilot planning system - email/phone call from vessel, agent, shipowner etc. - VHF communication with vessel - AIS transmitted from vessel
Output	correct vessel information
Precondition	pre-information received pre-information correct
Resource	not defined
Control	confirmation of vessel information by crew/pilot on board, based on: - visual observation of load lines - measurement from draught sensor confirmation by VTSO by checking MarineTraffic website (unauthorised)
Time	correct information should be received before vessel: - enters VTS/pilotage area - commits to route with draught restrictions

Various sources of variability or uncertainty may be seen. Draught information is received from several sources, which appear to conflict. Even the draught sensor on board the vessel, usually a control or check on the information received (function *'monitor vessel instruments'*), is not considered reliable. Once on board, the pilot is unable to confirm with certainty the correct vessel draught. Note that although the draught may itself be variable, it is *knowledge of the correct draught* which is uncertain in this case. We will not characterise it in terms of endogenous/exogenous or precision/time variability, but simply follow its effects on other functions.

6.2 AGGREGATION OF VARIABILTY

6.2. (a) CHARLIE enters VTS area

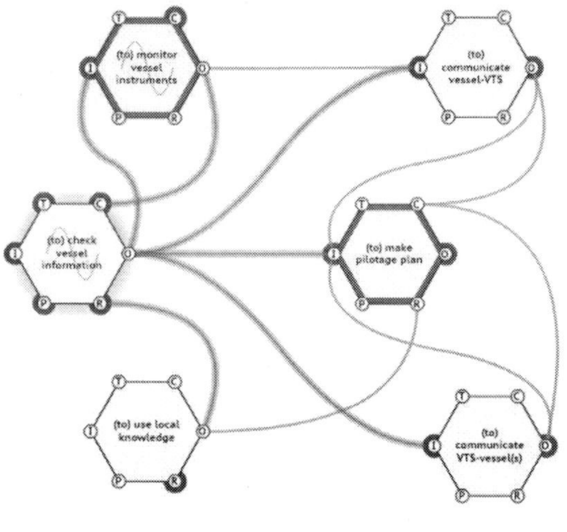

Figure 5a: FRAM instantiation of CHARLIE entering VTS area.

The uncertainty regarding the draught (functions *'check vessel information'* and *'monitor vessel instruments'*) affects the route the vessel should take (*'make pilotage plan'*) (Fig. 5a,b). This is communicated between vessel/pilot and VTS (*'communicate vessel-VTS'* and *'communicate VTS-vessel'*) when entering the VTS area, where uncertainty is managed by treating the vessel as a deep draught vessel (*'use local knowledge'*, including sub-functions *'know local geography'*, *'know local traffic patterns'*, *'have shiphandling experience'*, *'have pilotage/VTS experience'*) and agreeing that it will take the deep water route (*'make pilotage plan'*). Note that *'local knowledge'* has become a *resource* and *'communication'* has become a *control* aspect i.e. it is used to manage variability.

6.2. (b) CHARLIE approaches deep water route

On approaching the entrance to the deep water route (*'make pilotage plan'*) (Fig. 6a,b), a potential close quarters situation is foreseen by the VTS operator and pilot separately (both: *'check traffic situation'*). Due to the presence of ALPHA and BETA, CHARLIE will have difficulty entering the deep water route, unless one or more vessels deviates from their present course (foresight functions *'monitor traffic situation'*, *'monitor vessel motion'*). There is good visibility so the vessels should see each other visually (*'monitor weather conditions'*) and on their navigation displays (*'monitor vessel instruments'*). However, ship-to-ship communication is necessary to resolve the situation (*'communicate vessel-other vessels'*). The VTS operator is listening in to ensure this is done (*'monitor radio communication'*).

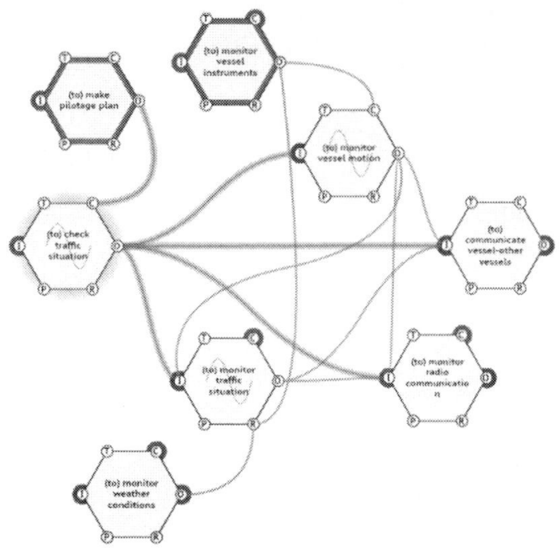

Figure 6a: FRAM instantiation of CHARLIE approaching deep water route.

6.3 MANAGING VARIABILTY

6.3. (a) CHARLIE negotiates safe passage

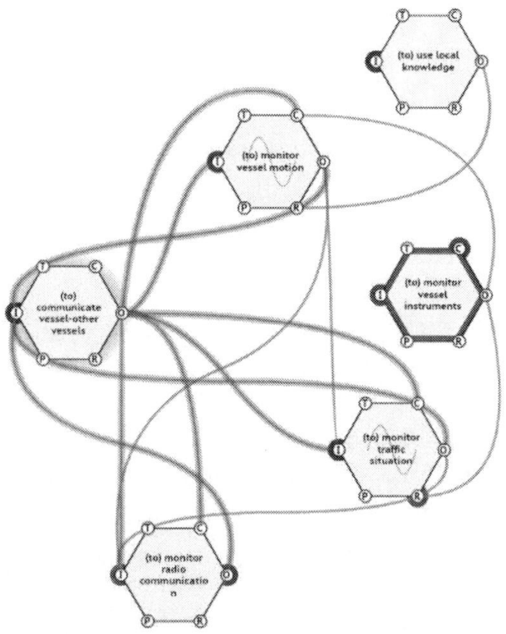

Figure 7a: FRAM instantiation of CHARLIE negotiating safe passage.

Firstly, CHARLIE negotiates with ALPHA to leave more space for CHARLIE to safely overtake (*'communicate vessel-other vessels'*, output: *'intended vessel movements'*) and keep clear of shallow water (*'use local knowledge'*) (Fig. 7a,b). This is monitored by pilot, VTS operator and vessels to ensure it has the desired effect (it becomes an input to and control aspect of *'monitor vessel motion'*, *'monitor traffic situation'*, *'monitor radio communication'*).

CHARLIE is still in a head-on situation with BETA (*'monitor traffic situation'*) and cannot turn to starboard due to shallow water (*'monitor vessel motion'*, *'monitor vessel instruments'*). Once again, the vessels negotiate (*'communicate vessel-other vessels'*), agreeing to disregard COLREGS Rule 14 and meet starboard-to-starboard; they then monitor until they have safely passed each other (*'monitor vessel motion'*, *'monitor traffic situation'*, *'monitor radio communication'*). The VTS operator does not interfere, as he trusts the vessels are in control of the situation (*'communicate vessel-other vessels'*, *'monitor radio communication'*, output: *'trust'*) but continues to monitor.

7. DISCUSSION

The case described above is a typical example of the work of pilots and VTS operators. (Incidentally, when the author at a later date asked the pilot about this case, he couldn't recall the actual event, but said, *"That's normal, it happens all the time"*.) It shows navigational assistance to be a cooperative practice between vessels and shore, not just between pilot, crew and VTS operator, but also between vessels themselves, instruments, geographical location, water depth and many other factors. It demonstrates how uncertainty or variability is a normal feature, present in various forms, but with good communication and trust between the parties, it can be managed and a potentially dangerous situation handled successfully. Although some navigational incidents do occur, it is not by divine intervention or chance that many are avoided, as Pepys believed, but that safety of navigation is improved by the work of skilled pilots, VTS operators and vessel navigators.

By describing this work in terms of functions (e.g. *'check vessel information'*, Table 2), one can clearly see the sociotechnical nature of navigational assistance. Here though, the author would depart from the usual classification of functions as human, technological or organisational, as described in [9, 10], and choose not to distinguish between them [11, 18]. For example *'check vessel information'* is performed by the pilot or VTS operator acting *together* with some or all of the vessel crew, vessel itself, draught sensors, chart display, emails and so on. While superficially it may seem a human function, it may be more accurate to say that, when performed, it becomes a sociotechnical function.

Pilots and VTS operators themselves do not distinguish between human and non-humans in their work, as can be seen from the examples in Section 5: *"we had two vessels coming southbound"*; *"if the vessels, on the bridge, are not worried..."*; *"this is BETA"*; *"BETA is local"*. Unless the topic of discussion is a particular individual, e.g. the pilot in the incident story, the *vessel* is the normal unit of interaction. (This is similar to the levels of abstraction in cognitive systems [20, 21] described in maritime traffic management by [8].) Therefore "following the vessel" is a valuable tool in understanding how work is performed, and may be used both as an observation strategy and as the focus for a FRAM analysis, as was hopefully demonstrated in this paper.

However, FRAM's notion of variability - as endogenous/exogenous, varying in precision/time [9, 10] - is dependent on whether a function is characterised as human, technological or organisational. Technological functions are seen as relatively stable, while human and organisational functions are more likely to be subject to variability [9, 10] By describing the function itself as sociotechnical and employing a more "fuzzy" concept of uncertainty rather than variability (see section 2), one can better follow the effects the uncertainty about the vessel draught has on the situation. The reader should note that uncertainty is not actually reduced or resolved, it is simply accepted by the actors as one of the conditions which shape performance in this case; it influences the choice of actions taken [11].

Using instantiations of a FRAM model in addition to a narrative-based description, one can visualise the interactions which take place, thereby literally "making work visible" [13]. However, the intricate web of hexagons and lines which are produced may be too complicated to be presented in a paper or article. This was resolved here by "zooming in" to key events in the case, making each individual interaction more visible, but potentially losing an overview of the whole situation. (Were the FRAM to be used in discussions between stakeholders in the design of new work systems, as envisaged by Hollnagel et al. [10], a better low-tech alternative may be to take a printed copy of the system model and draw the connections with a pen.)

This highlights one of the features which enables FRAM to describe "work as done", and potentially distinguishes it from many other systems approaches: the ability to move between a generic description of a work practice using a FRAM model, and a detailed exposition of how work is performed using instantiations of that model. This flexibility does produce some tension with traditional sociotechnical systems thinking, in which a system, its human-technological-organisational components and its boundaries are usually pre-defined - identified in [38] as a limitation of "harder" systems views - but opens for a "softer" approach in which STS/WS thinking may play a part.

For example, navigational assistance has certain central components - vessels, pilots, VTS etc. - and common features - communication, trust, use of local knowledge, preparation, foresight etc. - which may be understood and modelled as a sociotechnical system (Fig. 1a,b) [see also 8, 29, 37]. However, in the actual performance of work, as illustrated by the case study, some or all of these elements may come into play and interact with each other; the system reconfigures in time and space to adapt to the circumstances [see e.g. 12], allowing uncertainty to spread, but also to be managed. "Work as done" in a particular case may be seen as one of all possible configurations of the work system. But how this occurs is not fixed or pre-defined; it is understood by following the interactions and their effects, and the "system" it describes is a result of these interactions [see also 16]. One may therefore say that in performing a FRAM analysis and creating an instantiation of a FRAM model, one is tracing the formation of a network [11, 18]. The point however is not whether we define navigational assistance as a system or network, or both, but that there is added value in borrowing from different traditions when trying to understand and describe a real world practice [16, 25, 38].

8. CONCLUSIONS AND FUTURE WORK

The aim of this paper was to understand how navigational assistance is performed in everyday operations, and how pilots and VTS operators deal with uncertainty and variability in their work, thereby improving safety at sea. It is hoped that the approach taken here has provided a suitable framework to do so. The author envisions that in the future this approach may go beyond describing work, to being used as a tool for discussion in the design of new work systems [see also 10].

9. ACKNOWLEDGEMENTS

The author wishes to thank the participants of the study, the MonaLisa2.0 project for funding the field visits, and also Barbara Czarniawska, Torgeir Haavik and Gesa Praetorius for their inspiration and advice.

10. REFERENCES

1. IMPA, *'IMPA (International Maritime Pilots' Association) on Pilotage'*, 2014.

2. IMO, *'Pilotage'*, 2016.

3. IALA, *'Vessel Traffic Services Manual, Edition 6'*, 2016.

4. IMO, *'IMO Resolution A.857(20) Guidelines for Vessel Traffic Services'*, 1997.

5. GRECH, M., HORBERRY, T. and KOESTER, T., *'Human Factors in the Maritime Domain'*, 2008.

6. IMO, *'Proceedings of the Sub-Committee on Navigation, Communications and Search and Rescue (NCSR), 1st session, 30 June to 4 July 2014'*, 2014.

7. LIND, M., HÄGG, M., SIWE, U. and HARALDSON, S., 'Sea traffic management - beneficial for all maritime stakeholders', *Transportation Research Procedia*, 2016.

8. VAN WESTRENEN, F. and PRAETORIUS, G., 'Maritime traffic management: a need for central coordination?', *Cognition, Technology and Work*, 2012.

9. HOLLNAGEL, E., *'FRAM - The Functional Resonance Analysis Method: Modelling Complex Socio-technical Systems'*, 2012.

10. HOLLNAGEL, E., HOUNSGAARD, J. and COLLIGAN, L., *'FRAM – the Functional Resonance Analysis Method: A handbook for the practical use of the method'*, 2014.

11. LATOUR, B., *'Reassembling the Social: An Introduction to Actor-Network Theory'*, 2005.

12. SUCHMAN, L., 'Centers of Coordination: A Case and Some Themes', *Discourse, tools, and reasoning: essays on situated cognition*, 1993.

13. SUCHMAN, L., 'Making work visible', *Communications of the ACM*, 1995.

14. HUTCHINS, E., 'How a cockpit remembers its speeds', *Cognitive Science*, 1995.

15. HUTCHINS, E., *'Cognition in the wild'*, 1995.

16. HAAVIK, T. K., 'On components and relations in sociotechnical systems', *Journal of Contingencies and Crisis Management*, 2011.

17. LE COZE, J.-C., 'New models for new times. An anti-dualist move', *Safety Science*, 2013.

18. CZARNIAWSKA, B., *'Social science research: from field to desk'*, 2014.

19. KAGHAN, W. N. and BOWKER, G. C., 'Out of machine age?: complexity, sociotechnical systems and actor network theory', *Journal of Engineering and Technology Management*, 2001.

20. RASMUSSEN, J., 'Risk management in a dynamic society: a modelling problem', *Safety Science*, 1997.

21. HOLLNAGEL, E. and WOODS, D. D., *'Joint Cognitive Systems: An Introduction to Cognitive Systems Engineering'*, 2005.

22. HOLLNAGEL, E., WOODS, D. D. and LEVESON, N. C., *'Resilience engineering: Concepts and precepts'*, 2006.

23. HALE, A. and BORYS, D., 'Working to rule, or working safely? Part 1: A state of the art review', *Safety Science*, 2013.

24. HOLLNAGEL, E., *'Safety-I and Safety-II: The past and future of safety management'*, 2014.

25. HAAVIK, T. K., ANTONSEN, S., ROSNESS, R. and HALE, A., 'HRO and RE: A pragmatic perspective', *Safety Science*, 2016 (in press).

26. BEDFORD, T. and COOKE, R., *'Probabilistic Risk Analysis: Foundations and Methods'*, 2001.

27. HERRERA, I. A. and WOLTJER, R., 'Comparing a multi-linear (STEP) and systemic (FRAM) method for accident analysis', *Reliability Engineering and System Safety*, 2010.

28. CARVALHO, P. V. R. d., 'The use of Functional Resonance Analysis Method (FRAM) in a mid-air collision to understand some characteristics of the air traffic management system resilience', *Reliability Engineering and System Safety*, 2011.

29. PRAETORIUS, G., HOLLNAGEL, E. and DAHLMAN, J., 'Modelling Vessel Traffic Service to understand resilience in everyday operations', *Reliability Engineering and System Safety*, 2015.

30. CABRERA AGUILERA, M. V., BASTOS DA FONSECA, B., FERRIS, T. K., VIDAL, M. C. R. and CARVALHO, P. V. R. d., 'Modelling performance variabilities in oil spill response to improve system resilience', *Journal of Loss Prevention in the Process Industries*, 2016.

31. DE VRIES, L., 'Work as done? Understanding the practice of sociotechnical work in the maritime domain', Submitted for publication in *International Journal of Human Factors and Ergonomics*, 2016 (forthcoming).

32. DE VRIES, L., 'Success factors for navigational assistance: a complementary ship-shore perspective', *Proceedings of the Human Factors and Ergonomics Society Europe Chapter 2014 Annual Conference*, 2015.

33. STANTON, N. A., SALMON, P. M., WALKER, G. H., BABER, C. and JENKINS, D. P., *'Human Factors Methods: A Practical Guide for Engineering and Design, 2nd ed.'*, 2013.

34. GLASER, B. and STRAUSS, A. L., *'The discovery of grounded theory: Strategies for qualitative research'*, 1967.

35. CHARMAZ, K., *'Constructing grounded theory, 2nd ed.'*, 2006/2014.

36. CORBIN, J. M. and STRAUSS, A. L., *'Basics of Qualitative Research: Techniques and procedures for developing grounded theory, 3rd ed.'*, 2008.

37. MIKKERS, M., HENRIQSON, E. and DEKKER, S., 'Managing Multiple and Conflicting Goals in Dynamic and Complex Situations: Exploring the Practical Field of Maritime Pilots', *Journal of Maritime Research*, 2012.

38. CHECKLAND, P. B., 'Soft systems methodology: a 30 year retrospective', *Systems research*, 2000.

39. IMO, *'Convention on the International Regulations for Preventing Collisions at Sea'*, 1972.

11. AUTHORS BIOGRAPHY

Linda de Vries holds the current position of Project Coordinator and PhD candidate within the Maritime Human Factors and Navigation research group at Chalmers University of Technology. Her research focuses on safety in sociotechnical systems in a dynamic physical environment. She is a former sea captain and hydrographer/oceanographer, and has worked mainly in the offshore oil and renewable energy sectors.

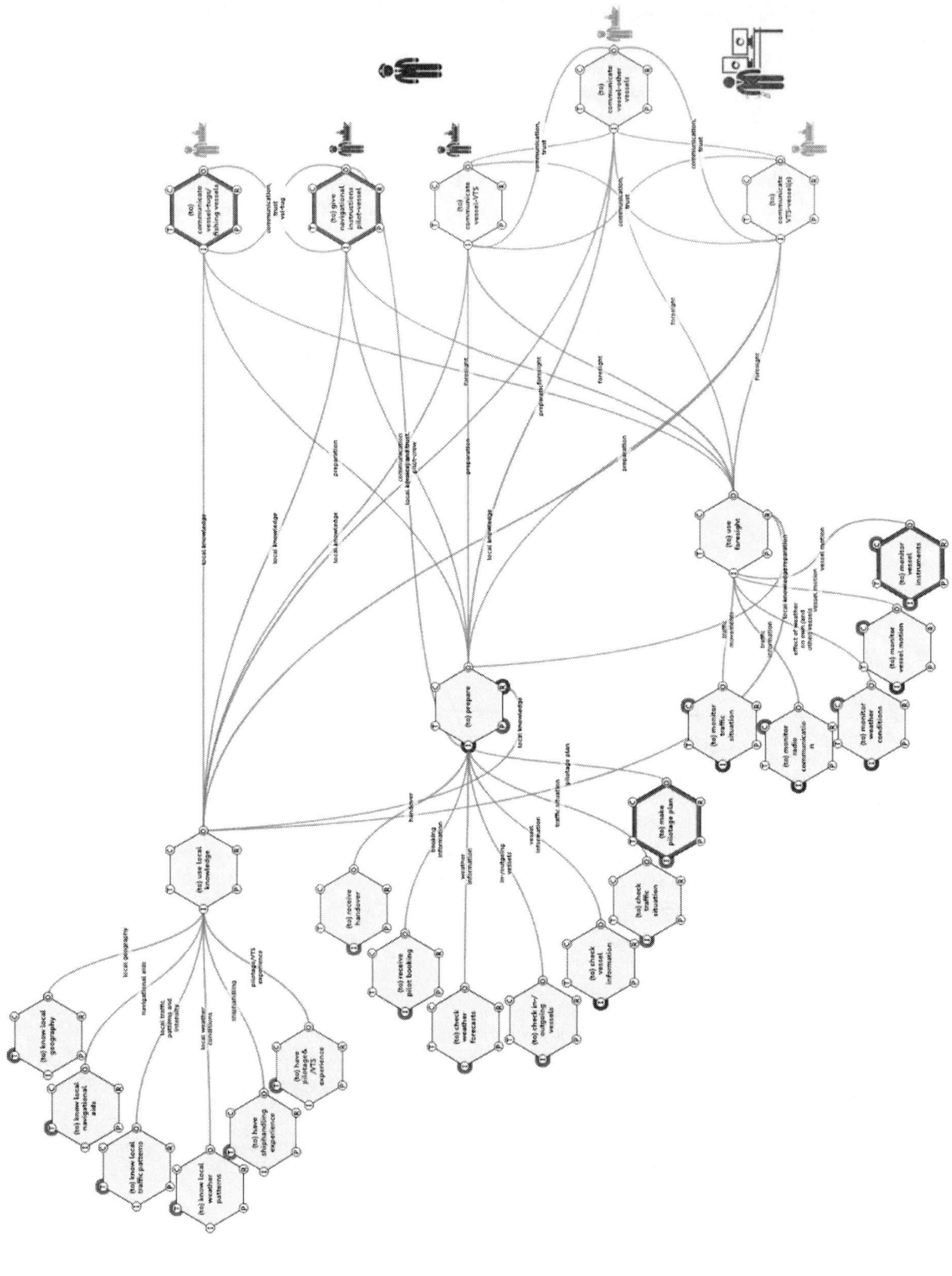

Figure 1b: Generic FRAM model of navigational assistance.

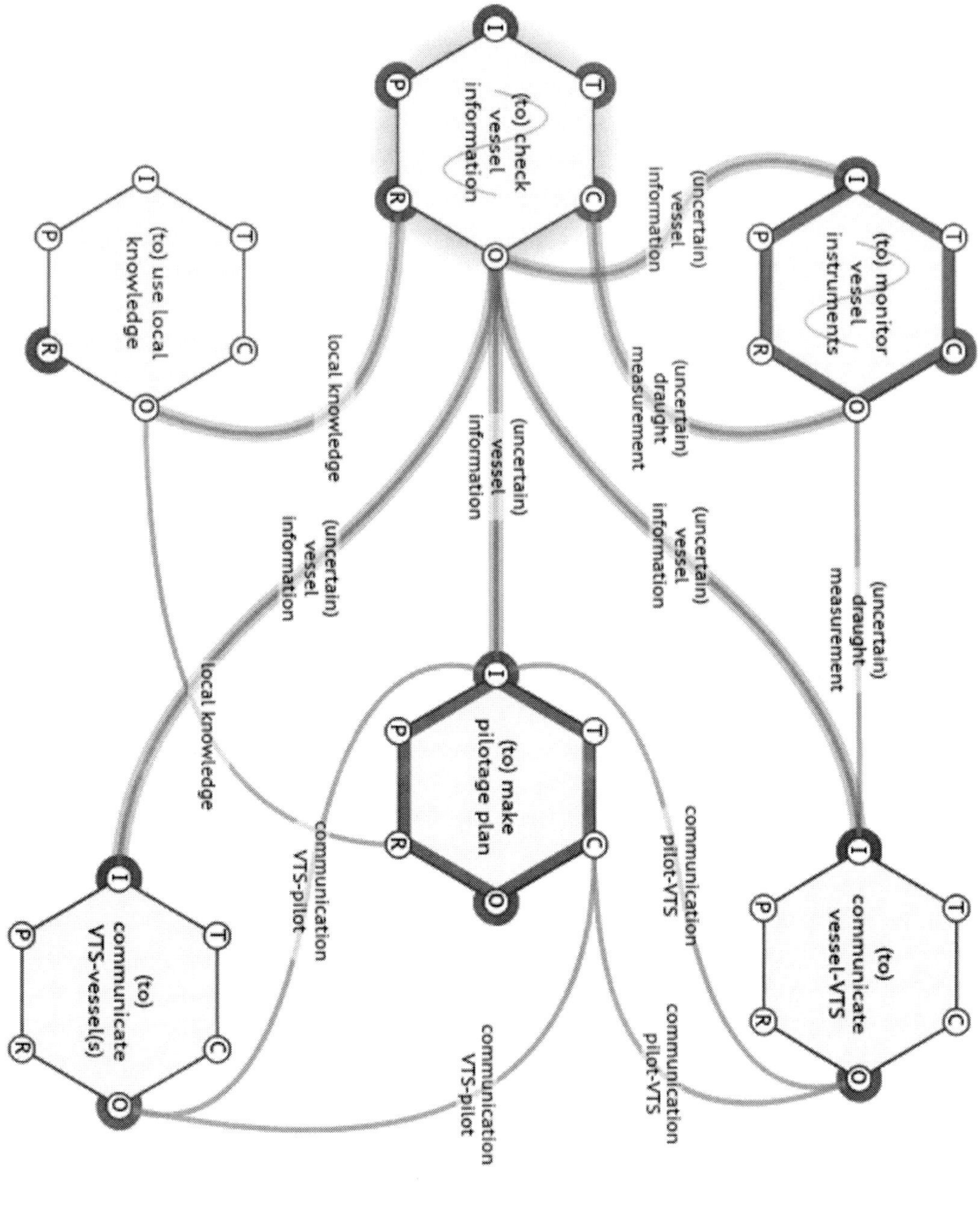

Figure 5b: FRAM instantiation of CHARLIE entering VTS area.

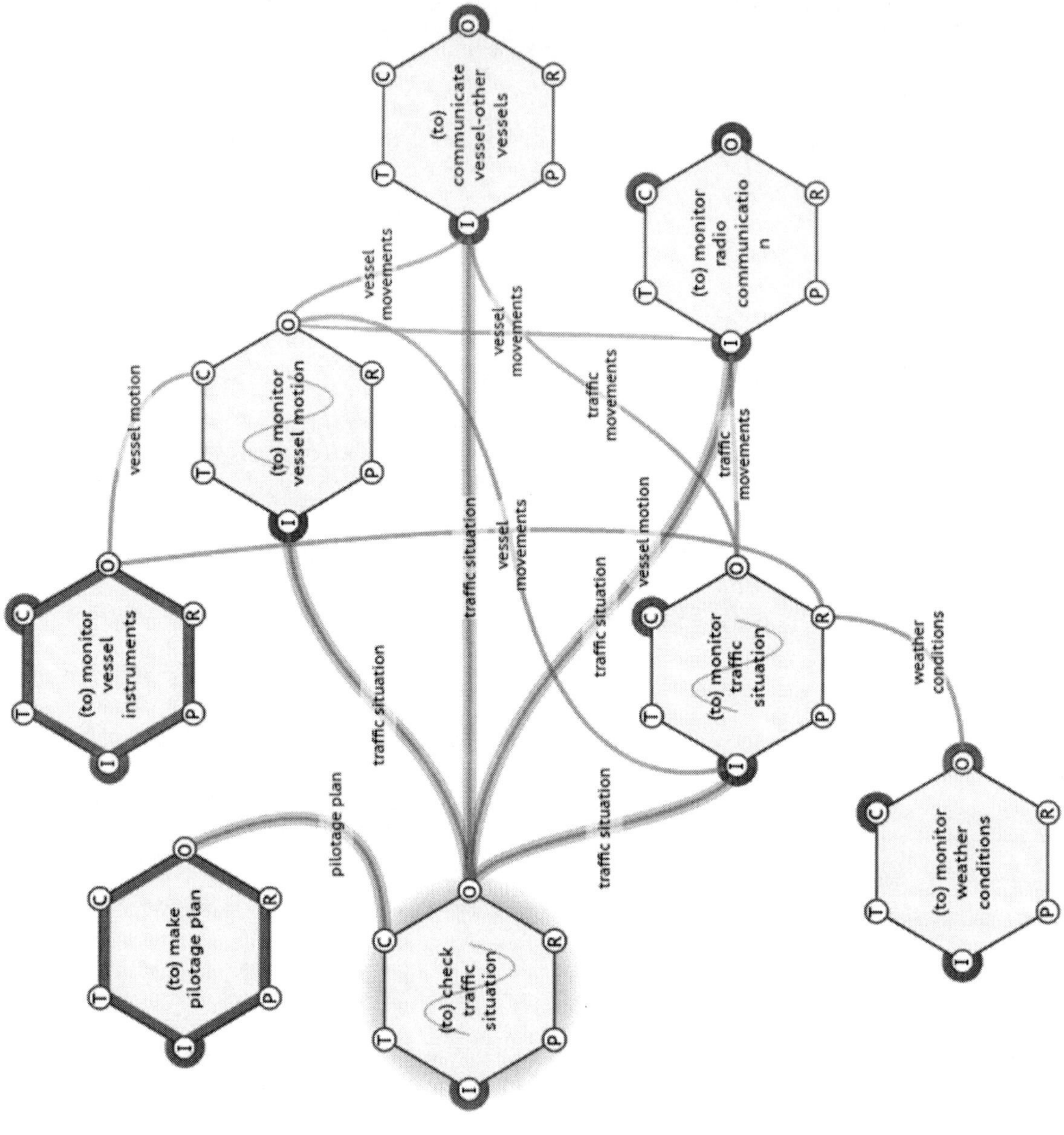

Figure 6b: FRAM instantiation of CHARLIE approaching deep water route.

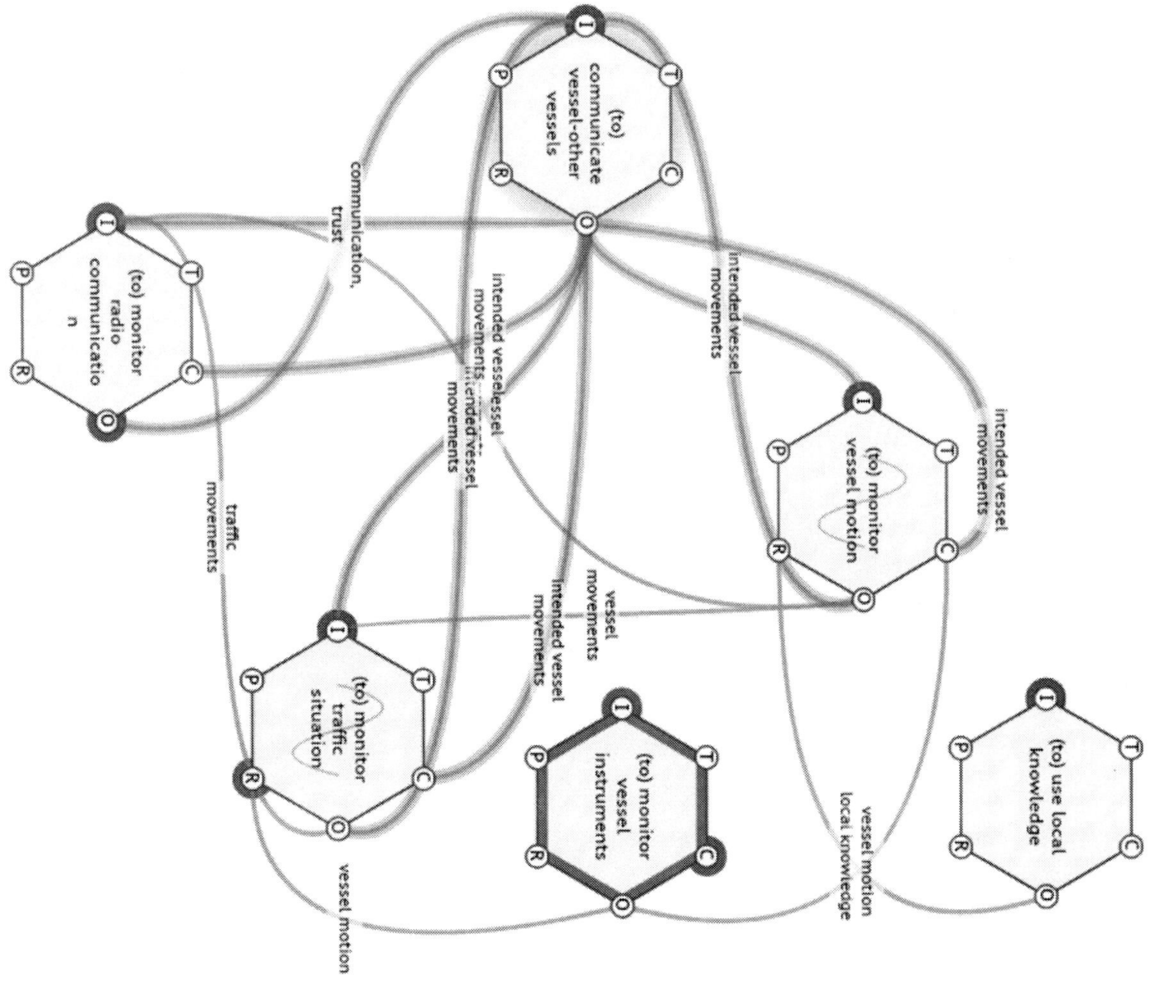

Figure 7b: FRAM instantiation of CHARLIE negotiating safe passage.

STANDARDISED INFORMATION ARCHITECTURE TO SUPPORT THE DYNAMIC NAVIGATION (DYNAV) STANDARD OPERATING PROCEDURE

T Dobbins, STResearch, UK
J Hill, Trident Marine, UK
T Brand, Strongwake, CANADA
T Thompson, T-Design, UK
S McCartan, EBDIG-IRC, NTDC, Coventry University, UK

SUMMARY

Fast craft operations are typically undertaken at high tempo and in congested littoral areas. Operating at the edge of the operational envelope places the crew under extreme stress to deliver the required performance whilst managing the inherent risks. The crew operates as part of a Joint Cognitive System (JCS), the interaction of the crew and technical components working together within the harsh environment. The adoption of the DYnamic NAVigation (DYNAV) methodology has highlighted the need for systems to support this SOP. Information is fundamental to developing and maintaining SA, it must be formatted to specifically support DYNAV and the crew's ability to make safe and effective decisions. This requires the display design development to use a Human Centred Design (HCD) approach. This HCD approach also addresses the control of the information, which must be simple, intuitive and effective for operations in a harsh RS & WBV environment. Although new technology and systems are continually being developed, e.g. Head-Up Displays and automation, they must still support the decisions that the crew make. This paper delineates the standardised information architecture requirements for DYNAV.

1. INTRODUCTION

Speed is relative. Travelling at >50KTS in the open ocean, on smooth deep water, is a simple command and control task for the crew. The same craft, travelling at the same speed, but in shallow congested water requires great skill from the crew, operating within an effectively designed system, to maintain an appropriate level of safety and performance. It is therefore pointless to develop crew support system for simple, open ocean transits. Rather, the system's features must support the crew for their worst-case scenarios at the edge of the operating envelope [1]. Worst-case operating scenarios are not found in the leisure sector, they are found in naval, military and Search & Rescue (SAR) operations. Therefore systems principally designed for the leisure sector are NOT fit-for-purpose.

Naval, military and SAR operations reply on sophisticated Command & Control (C2) systems to successfully complete their tasks. All of these tasks are underpinned by a foundation of geo-location / navigation capability. The failure of navigation systems to successfully support the crew operating at higher speeds in poor sea conditions has been recognised by experienced operators for a long time, and unfortunately has not been commonly addressed by the system designers, manufacturers, and suppliers – even though outline guidance [2] has been provided.

Therefore to support the crew's demand for fit-for-purpose navigation system a standardised information architecture / layout has been developed to specifically support the internationally recognised DYnamic NAVigation (DYNAV) Standard Operating Procedure (SOP) [3].

2. OPERATING ENVELOPES

Operating at the edge of the operational envelope puts the crew under extreme stress to deliver the required performance whilst managing the inherent risks and possible disastrous outcomes. An example of a generic HSC operational envelope is shown in Figure 1.

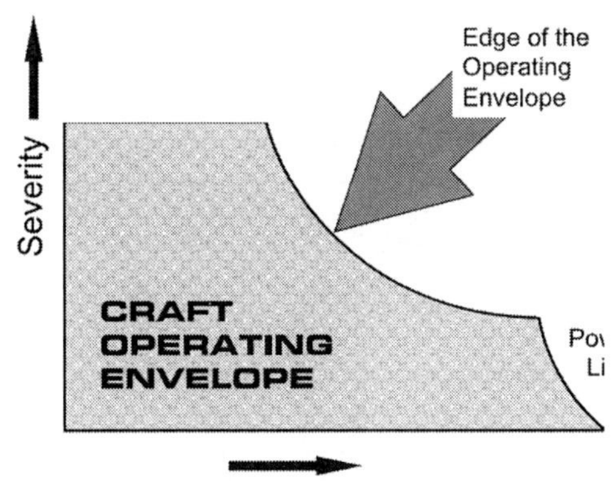

Figure 1: Example of a Generic High Speed Craft (HSC) Operational Envelope.

The current displays, designed for leisure applications, may have limited application over a range of speeds but at a minimal sea state, i.e. in smooth sea conditions with minimal exposure to Repeat Shock (RS) and Whole Body Vibration (WBV) exposure.

The operational requirement of the craft is to operate at the edge of it's design envelope. Unfortunately the poor design of the C2 interface system does not allow the crew to operate the craft safely and effectively at the edge of its operational envelope. Therefore the effective operating envelope of the craft is regarded as illustrated in Figure 2.

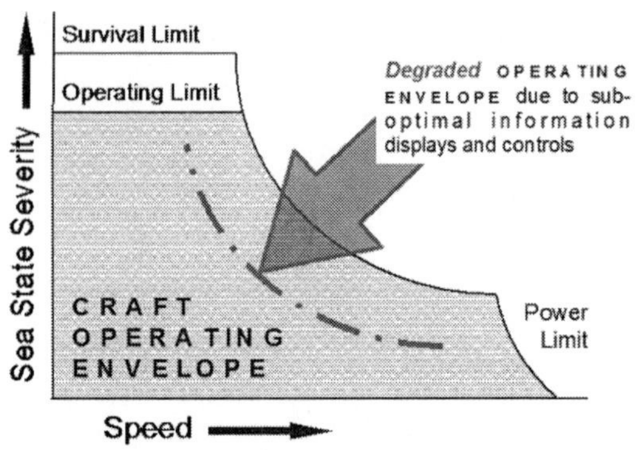

Figure 2: An Illustration of how a High Speed Craft (HSC) Operational Envelope is Degraded by Sub-Optimal Information displays.

3. JOINT COGNITIVE SYSTEM

The crew operates within the operating envelope as part as part of a Joint Cognitive System (JCS), the interaction of the crew and technical components working together within the harsh environment – as illustrated in Figure 3.

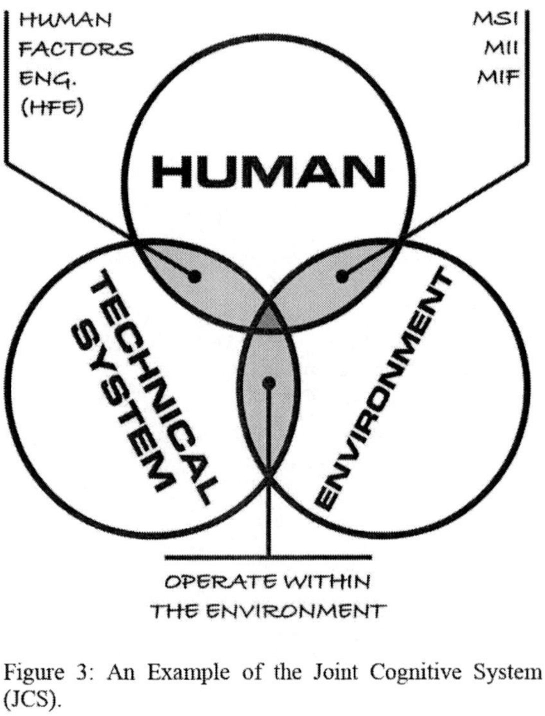

Figure 3: An Example of the Joint Cognitive System (JCS).

Many operations, e.g. SAR and patrolling, rely on navigation as the underpinning capability. For example the craft information system may use overlapping layers of information (see Figure 4) to build up the complete operating picture. It is essential that the order of the layers is standardised to support interoperability.

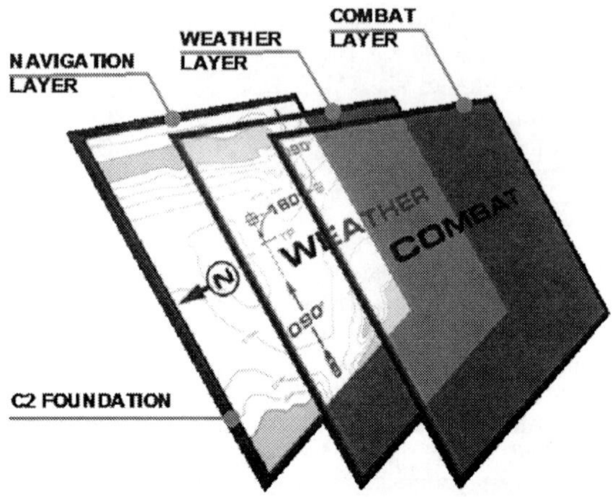

Figure 4: An Example of How the Navigation Display Provides the Foundation for Subsequent Tasks and Command & Control (C2).

The craft's geospatial position within it's environment underpins all of the other layers. With a multi-layer information display system it is simple to overburden the individual / team with data / information, much of which will be of no use to them ! Therefore there is a need to both de-clutter the display(s) and focus on displaying the right information at the right time.

To ensure the right information is displayed at the right time it is essential to optimise the foundation navigation layer, based on the information required by the navigator, as defined by the DYNAV SOP.

4. THE DYNAV SOP

The development [4] and adoption of the DYNAV methodology [3] by the international community has highlighted the need for systems to support the SOP. The DYNAV methodology bases each turn on a standard series of phases; *Plan – Communicate – Execute – Control*, as shown in Figure 5.

Figure 5: The Phases of the DYnamic Navigation (DYNAV) Methodology.

During the Plan phase the system must display specific information, dictated by the standard DYNAV communication protocol, to the Navigator. At higher speeds, and in dynamic environments the Plan phase can be relatively short and therefore the information must be presented intuitively and in a format where it can be rapidly perceived and assimilated.

5. DEVELOPING THE DYNAV INFORMATION ARCHITECTURE

The requirement for an enhanced information architecture has previously been descried [5] along with the need to engage in a Human Centred Design (HCD) Process (see Figure 6) – which is recognised by the IMO [6]. The objective is to support the process (see Figure 7) for the development of a DYNAV specific display.

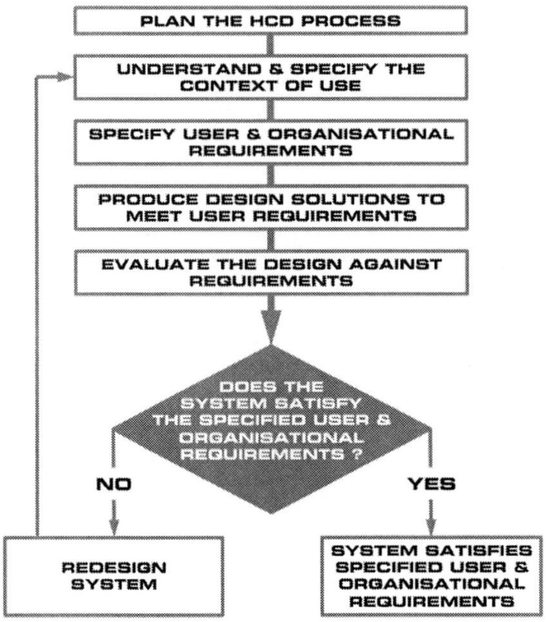

Figure 6: The Human Centred Design (HCD) Process.

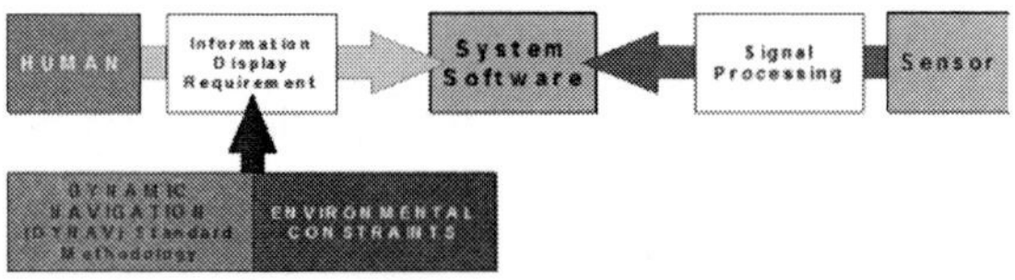

Figure 7: The DYnamic NAVigation (DYNAV) Information Display Development Process.

The DYNAV display development process starts with understanding the role / tasks of the Navigator and the subsequent decisions that are made. This decision making process requires the Navigator to have an appropriate level of Situation Awareness.

5.1 SITUATION AWARENESS AND DECISION MAKING

Situation Awareness (SA) is the foundation of the Navigators decision-making process within the DYNAV methodology that operates as part of the JCS. Within the context of DYNAV it has described as, "*an appropriate level of understanding of the craft's situation within its environment to make safe judgments and decisions about its current and future navigation*" [3], and requirements to support SA are described within the HSC HFE Design Guide and Supplement [2].

Information is fundamental to developing and maintaining SA, for DYNAV this requires the crew to make sense of both the external view and the information provided by the systems displays. The way that the Navigator uses their SA to make decisions include Recognition Primed Decision making (RPD) and what may be described as intuition. These decision-making processes are further described within the DYNAV Manual [3].

5.2 CREW VISION: EXTERNAL & INTERNAL

As previously stated, current displays and display logic do not, typically, effectively support the crew and degrade their ability to operate safely at the edge of the operational envelope (refer to Figure 2). The following requirements must be included:

- The external view must be uninterrupted, or have minimal interruption to the required sight lines.
- The internal displays must be placed where the crew can see them with the minimum of eye and head movement.
- The display must be located so the information directly relates to the external view and directions are coincident.
- It must be possible for the crew to intuitively control the display and it's information with the minimum of actions / menu levels.

The required position of the displays is covered in the HFE Design Guide and it's Supplement [2] and an example illustrated in Figures 8 and 9.

Figure 8: An Example of the DYnamic NAVigation (DYNAV) Display Console Layout Within the High Speed Craft (HSC).

Figure 9: An Example of the DYNAV Display Consoles Position Within the High Speed Craft (HSC) to facilitate both an effective external view and a direct view of the Internal displays.

5.3 DYNAV INFORMATION ARCHITECTURE

The information must be formatted to specifically support the DYNAV methodology and the crew's ability to make safe and effective decisions.

Current display technology means that the crew can have a large 'glass cockpit', giving them the opportunity to start with a blank-sheet-of-paper. Previously designers may have been constrained by the size and cost of the displays, this is no longer the case.

Different crewmembers have different roles / tasks and so require information to the displayed / formatted with different architecture. These specific roles include the Navigator, Driver and Commander.

5.3(a) Navigator Information

The following display characteristics / features are required to support the Navigator's task:

- Focus on the future course of the craft, i.e. subsequent turns.
- Display orientation – portrait to increase the 'view' in front of the craft
- Text size – increase so that the words / numbers can be easily read in a high shock and vibration environment
- Declutter, i.e. remove all unnecessary features to focus on the important information
- DYNAV communication requirements – ensure the display focuses on the essential, specific aspects of DYNAV:
  a. The direction of the next turn
  b. Turn point location
  c. Distance / time to next turn point
  d. Post-turn course / heading
  e. Post-turn head mark
  f. Potential dangers
  g. Max. and min. depths
  h. Available movement space

An example of how the Navigator's required information (re: points a - h above) may be initially structured is illustrated in Figure 10.

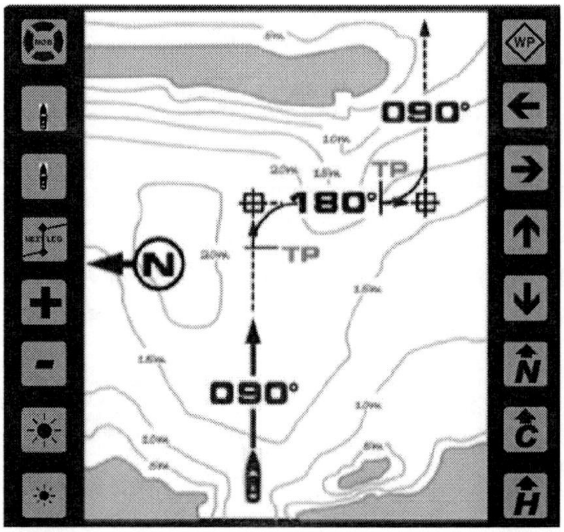

Figure 10: Example of the Standardised DYnamic NAVigation (DYNAV) Display for the Navigator highlighting features to support the standardised DYNAV communication protocol.

5.3(b) Driver Information

The Driver's display is similar to the Navigator's display but focuses on the environment directly in front of the craft in contrast to the Navigator who focuses on the future position of the craft. The example shown in Figure 11, highlights the details of the current and next leg, whilst Figure 10 – the Navigator display – illustrates the current and two future legs.

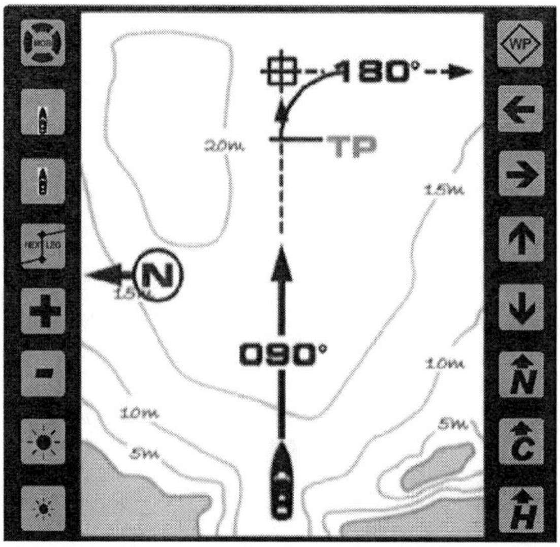

Figure 11: Example of the Standardised DYnamic NAVigation (DYNAV) Display for the Driver.

5.3(c) Commander Information

The boat Commander has a more strategic role than the more immediate / tactical roles of the Driver and Navigator. Therefore they are more involved in understanding the 'bigger-picture' of the craft's position within it's environment and the other vessels and features around it. The Commander will also be in contact other vessels and agencies. To facilitate a common frame-of reference and shared-SA the Commander will typically use a North-up chart orientation.

6. CONTROLLING INFORMATION

The HCD approach addresses the control of the information – as well as its display, this must be simple, intuitive and effective for operations in a harsh RS & WBV environment.

Control of the system via a touch screen interface may be popular and fashionable, but it should not be used for operating the system when underway as the high levels of RS & WBV make this an impossible interface to use safely and effectively.

Options to manipulate the screen and its information must include controls (e.g. buttons, switches, dials) that are large and rugged enough for the harsh marine environment. As outlined in the HFE Design Guide and it's Supplement [2]. Similarly the controls may be mounted on the seat (typically providing shock mitigation) and should be robust enough to cope with the motion environment and the postural stability requirements of the seat occupant. Examples of typical controls are shown in Figure 12 and 13.

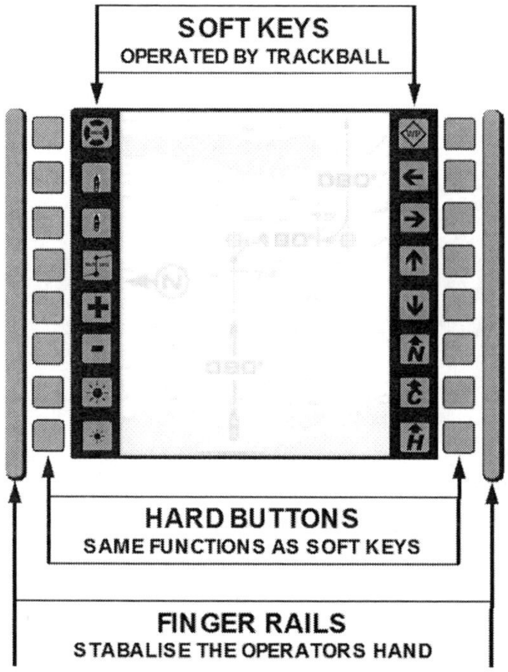

Figure 12: Example of Standardised Controls Integrated with the DYNAV Display on console.

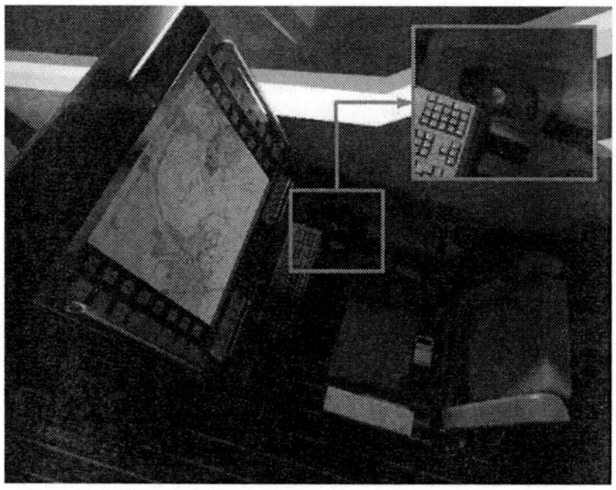

Figure 13: Example of a Trackball Located on the Suspension Seat Armrest to Control the DYNAV Display Soft Keys and other Functions.

The crew have many specific control requirements that must be actuated with a single control function, e.g. a single button push. The designer must understand these requirements, and not design a system that buries important functions within multiple menu layers, e.g. switching between North-up and course-up chart orientations. Examples of functions that require specific controls, i.e. their own single-function buttons, are illustrated in Figure 14 below. Other button functions may be required along with user-defined functions for specific applications.

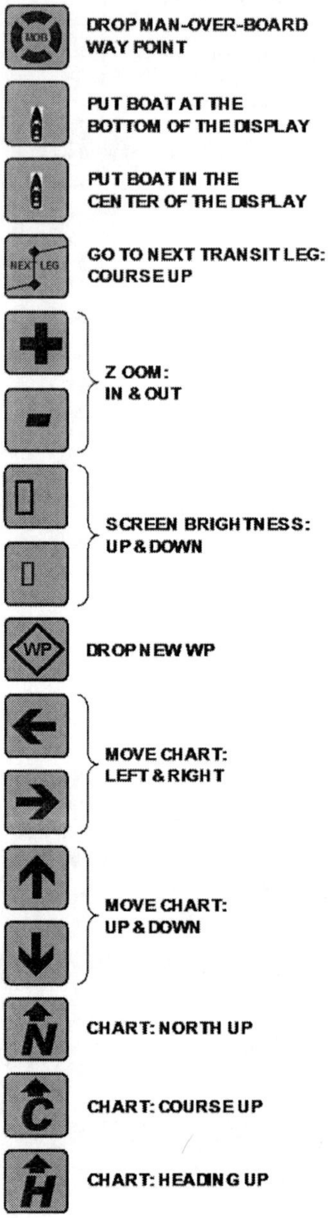

Figure 14: Examples of Single Function Control Button Functions supporting the DYnamic NAVigation (DYNAV) Methodology.

7. NEW DISPLAY SYSTEMS

Although new technology and systems are continually being developed, e.g. Head-Up Displays (HUD) and increasing levels of automation, they must still support the decisions that the crew makes.

7.1 HEAD-UP DISPLAY

A HUD typically provides specific, essential information within the crew-members field of view of the external environment. This may typically be projected onto a panel in front of the operator or the windscreen. The advantage of the HUD is that it stops the navigator / driver from having to look down for the information required, taking their eyes off the external view and potentially compromising the safety of the craft. An example of a HUD, designed to support the DYNAV SOP is shown in Figures 15 and 16.

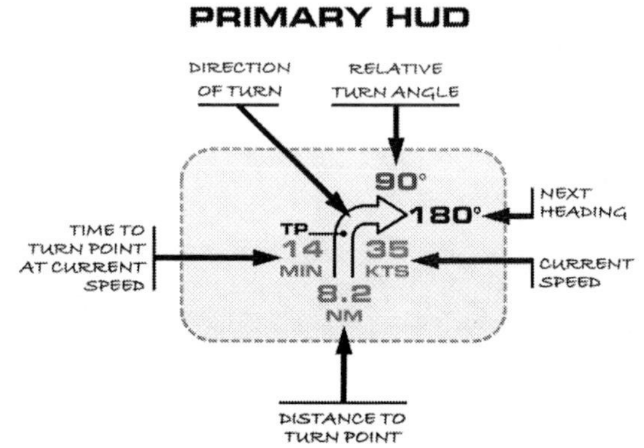

Figure 15: Example of a Standardised DYnamic NAVigation (DYNAV) Head Up Display (HUD) layout.

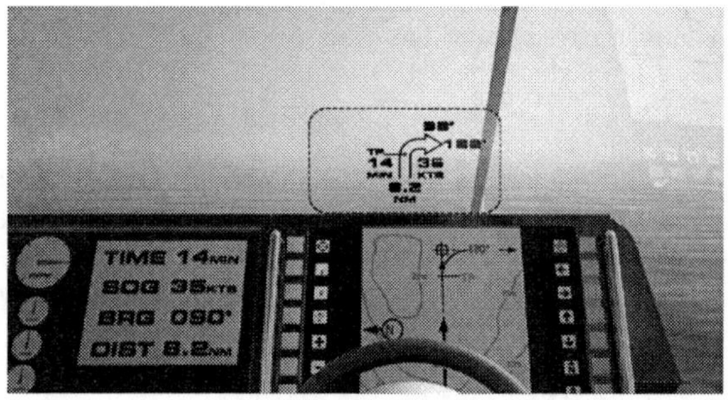

Figure 16: Example of the Standardised Head Up Display (HUD) Implemented in a High Speed Craft (HSC).

7.2 AUGMENTED REALITY

Augmented Reality (AR) is becoming increasingly common in computerised systems, particularly hand-held devices. The use of AR systems have been demonstrated before [7] and are of particular use in blind-pilotage (i.e. restricted external view) scenarios. An example being the use of infra-red camera displays with an overlay of the waypoints, turnpoints, routes, etc. This type of AR display typically requires an additional screen, therefore the craft's Human Factors Engineer / Designer must account for this when developing the cockpit layout.

8. CONCLUSIONS

The development of a standardised DYNAV information architecture supports crew performance, safety and interoperability. The effective implementation of new display systems, such as augmented reality and head-up displays, using a standardised architecture, offers significant potential to enhance the operational envelope of HSC.

This paper focuses on the need to support HSC navigation. The overall command and control of the craft includes other aspects such as steering, trim, autopilot, engine / systems monitoring, communications, etc. All of which require the same level of design focus to develop the total system to an effective level.

9. REFERENCES

1. Dobbins, T., Horton, J. and Scott, R. (2014) Defining Operational Envelopes To Support Human Systems Integration Within Fast Craft Design. Conference proceedings; RINA Human Factors Conference. London, UK.

2. High Speed Craft Human Factors Engineering Design Guide and Supplement. ABCD-TR-08-01 v1.0. www.highspeedcraft.org

3. DYNAV Manual. ISBN 978-1-5262-0347-2. www.dynav.org

4. Forsman, F., Dahlman, J. and Dobbins, T. (2011) Developing a Standard Methodology For Dynamic Navigation in the Littoral Environment. Conference Proceedings; RINA Human Factors in Ship Design Conference, London, November, 2011.

5. Dobbins, T., Forsman, F., Hill, J., Brand, T., Dahlman, J., Harris, D., Smoker, A., Stark J. and MacKinnon, S. (2013) Information Architecture for Fast Response Craft – Command & Control & Human Systems Integration. Conference Proceedings; RINA SURV-8 conference, Poole, UK.

6. Australia, E-NAVIGATION STRATEGY IMPLEMENTATION PLAN : Report of the Correspondence Group on Harmonization of Guidelines related to e-navigation 2015, International Maritime Organization: London.

7. Pecota, S., Grabowski, M. and Holder, E. (2015) Augmented Reality Navigation Displays: Maintaining situational awareness in the e-Navigation era. The Coast Guard Proceedings of the Marine Safety and Security Council. Vol. 72(2): pp 52-55.

10. AUTHORS BIOGRAPHY

Trevor Dobbins holds the current position of Director at STResearch Ltd. He has published widely on HSC human factors and is the principal author of the High Speed Craft Human Factors Engineering Design Guide. His previous experiences include projects for the UK MOD, RNLI, dstl, and the US Navy.

Jon Hill holds the current position of Director at Trident Marine Ltd. He retired from the Royal Marines as the Warrant Officer responsible for Landing Craft and high-speed boat training whilst also being responsible for craft and operational development. He currently works internationally providing training support to Governments/Coast Guards and advising on craft and training development.

Tyler Brand holds the current position of Director at Strongwake Ltd. He previously served as a Canadian Coast Guard Officer, including responsibility for fast craft training. He subsequently worked developing innovative training solutions including the use of simulation integrated with classroom and on-water training.

Tim Thompson holds the current position of part-time CAD lecturer in the Department of Industrial Design at Coventry University, UK. His CAD teaching is informed by his professional practice within his design consultancy, T-Design, which specialises in photorealistic visualisation for the superyacht design and interior design industries.

Sean McCartan holds the current position of Course Tutor, Boat Design at Coventry University, UK. His key research area is TOI (Transfer of Innovation) from other sectors to the marine industry, in the areas of Design-Driven Innovation (DDI), advanced visualisation and Human Systems Integration (HSI). He leads the EBDIG (European Boat Design Innovation Group) network, which includes Chalmers University; Genoa University; TU-Delft; and a number of leading European marine design consultancies.

Human Factors, 28-29 September 2016, London, UK

THE TRAINING SYSTEM: DEVELOPING CREW COMPETENCE WITHIN THE VIRTUAL AND REAL WORLDS

J Hill, Trident Marine Ltd, UK
T Brand, Strongwake, CANADA
T Dobbins, STResearch Ltd, UK

SUMMARY

Fast craft operations are high risk and push the crew to the edge of their operational capability in the full range of environmental conditions. Effective Command & Control in these conditions is essential if risk is to be effectively managed and safety maintained. This capability is founded on education, training, and experience. Defining the required competences facilitates both training and assessment, which are delivered via classroom, computer-based and on-water training. Competencies can be divided into three aspects; knowledge, skills and attitude (KSA). Classroom taught knowledge is essential, but it is the crew's demonstration of their skills and attitude, when put under pressure, on the water, that is of the ultimate competence test. To support the training and learning process, instructors use a range of tools from simple performance profiling to full mission simulation. The ability to practice Standard Operating Procedures (SOPs) prior to operations and in increasing demanding situations is vital to maintain capability. Simulation, correctly integrated into a training syllabus, is a valuable tool in the education progression of developing individual and team competence and proficiency. Simulation can also put the crews into increasingly more dangerous scenarios, gaining experience that is not available in real life without extreme dangers to life or craft. The essential aspect for the development of competence is the training curriculum, specifically focusing on simulation systems, that develops both individual and team competence.

1. INTRODUCTION

Fast craft operations are high risk and push the crew to the edge of their operational capability in the full range of environmental conditions. Effective Command & Control in these conditions is essential if risk is to be effectively managed and safety maintained. This capability is founded on education, training, and experience, but it is essential that the appropriate knowledge, skills and attitude (KSA) are imparted to deliver this capability. The issues of risk, competence, interoperability and qualifications for fast craft operations have previously been highlighted [1, 2]. Continuing developments and increasing system maturity within the simulation domain [3, 4] are now accelerating it's potential adoption by the fast craft community.

2. COMPETENCE DEFINITION

Defining the required competences facilitates both training and assessment, which are delivered via classroom, computer-based and on-water training. Competencies can be divided into three aspects; Knowledge, Skills and Attitude (KSA):

2.1 KNOWLEDGE, SKILL AND ATTITUDE

2.1 (a) Knowledge (KNOW WHAT); is a body of information applied to the performance of a function.

2.1 (b) Skill (KNOW HOW); is an observable competence to perform a learned psychomotor (conscious mental activity) act.

2.1 (c) Attitude (KNOW WHY); is the competence to perform an observable behavior that results in an observable product.

2.2 LEVELS OF COMPETENCE

One of the roles of the Training Instructor is to develop and maximise the competences of the craft Navigator and Driver. For the Trainees to develop their competence they follow an incremental process as they develop their KSAs:

a. Unconscious incompetence: The trainee typically does not understand or know how to do something and importantly they do not recognise that they don't know. It is also typical that they recognise the usefulness of the knowledge and skills they require. It is important that the trainee understand the value of the training to provide the intrinsic motivation to learn.

b. Conscious incompetence: The Trainee recognises that they have a lot to learn and the value of the knowledge and skills they wish to acquire. During this stage they will be making a lot of mistakes – but these are part of the learning process.

c. Conscious competence: The trainee is knowledgeable and has acquired the basic skills but they have to concentrate and focus to be successful.

d. Unconscious competence: No longer a trainee, the skills have become 'second nature'. Essentially the individual no longer has to concentrate and so has the ability to focus on more cues and develop more creative solutions, i.e. demonstrate the *Art of Navigation*.

The process of enhancing competence via training and subsequent experience is illustrated in Figure 1.

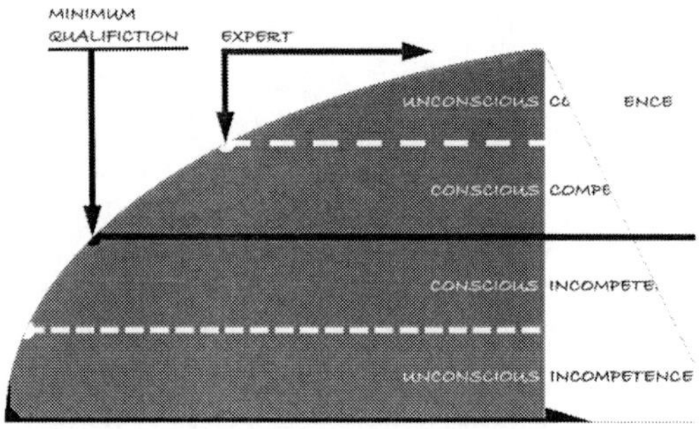

Figure 1: An Illustration of How the Individuals Level of Competence Increases with Training and Experience.

3. QUALIFICATIONS

Qualifications are an essential foundation of professional competence, particularly for tasks that have an inherent safety risk.

They are renewed at the appropriate interval, with the individual demonstrating evidence of Continued Professional Development (CPD), and KSAs maintained at the appropriate level. The demonstration of competence via qualification / certification provides an established benchmark against which comparisons can be made. Therefore qualifications provide organisations with a system of quality assurance and control. Where qualifications are recognised by multiple organisations / agencies / nations this standardisation provides the capability to facilitate interoperability and subsequently enhance operational effectiveness, particularly for joint-force operations. This interoperability requirement is highlighted by VAdm Parker of the USCG [5]:

New international training standards are required that address not just the competence of fast boat handling in the vicinity of other platforms, but the capability to be interoperable with other organizations / agencies during maritime operations.

Qualifications also provide the ability to rapidly disseminate best practice with an appropriate benchmark.

4. EDUCATION & TRAINING

The education & training process for developing competence has previously been a two stage process; classroom education followed by on-water training. Contemporary education and training is now a three stage process; classroom education, simulation training followed by on-water training.

4.1 CLASSROOM EDUCATION

Classroom taught Knowledge is essential, but it is the crew's demonstration of their Skills and Attitude, when put under pressure, on the water, that is the ultimate competence test

4.2 SIMULATION TRAINING

To support the learning and training process, instructors use a range of tools from simple performance profiling to full mission simulation. Simulation, correctly integrated into a training syllabus, is a valuable tool in the progression of developing individual and team competence and proficiency. Simulation can also put the crews into increasingly more dangerous scenarios, gaining experience that is not available in real life without extreme danger to life and craft.

4.2 (a) Simulation Integration

With the continued development of simulation the integration of these systems still, unfortunately, takes a back seat in their development. The correct integration of simulation into the training system reduces training costs, time, risks and increases individual and team competence. Fast craft simulation supports the delivery of crew competences, including the roles of the driver and navigator, using the same system.

4.2 (b) Simulation & Gaming

The difference between games and simulation is often blurred. During the last few years the use of gaming and simulation has merged with some organisations preferring a more entertaining approach to learning. In simple terms gaming requires little background storyline and is generally based in fantasy where as simulation requires more detailed real-life scenarios and facts to support situation analysis and proof of learning. There are more elements and differences between games and simulation including:

- Score Quality and Goal Completion
- Type of Feedback
- Point of View
- Interaction / Decision making
- Information or Rules

These points need to be taken into account when deciding how simulation is integrated into an organisation's training system / curriculum. The following are examples of simulation that fulfil different roles.

4.2 (c) PC-based Trainer

A PC-based simulator / trainer can vary from basic computer programmes with short specific training scenarios such as Pilotage, IRPCS or launch & recovery drills. An example of a system is shown in Figure 2. PC-based trainers are not limited to individual systems but can be networked together in-order to train multiple operators as a coordinated team in a shared virtual environment.

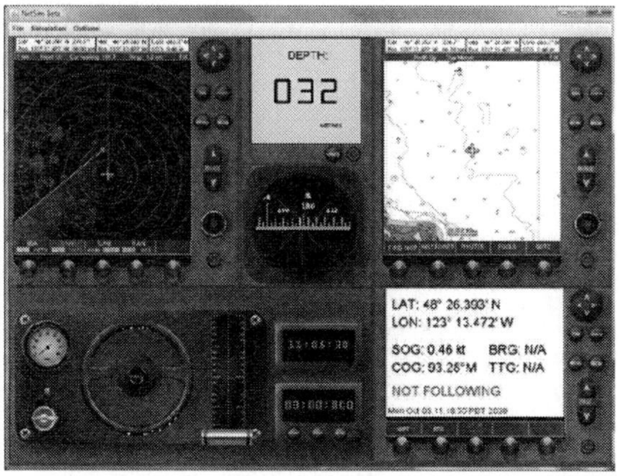

Figure 2: Example of a PC-Based Simulator / Trainer (*NETSim, VMT*)

4.2(d) Part Task Trainer (PTT)

A part task trainer (PTT) is a training device that is designed to train a member of the crew or maintenance staff on a particular task associated with the craft. The PTT is a cost-effective training solution that allows operators or maintenance personnel to familiarise themselves with a particular craft or system without having to use the actual craft.

4.2 (e) Fixed-Base Simulator

Fixed-base simulation has in the past taken a back seat to full-motion simulation, but in the last few years, and with the introduction of better immersive environments, organisations are taking and increasing interest in fixed-base systems. Previously fixed-based simulation was considered limited, but now - with a greater understanding of the required learning outcomes (e.g. role specific KSAs), they are now being recognised as potentially having a greater cost-benefit ratio than motion-based simulation.

4.2 (f) Motion-Base Simulator

Motion-based simulation has always been seen as the best option for closely recreating the operating environment and giving the best opportunity to gain the required knowledge / skills. However simulation within the high-speed craft sector is limited by the systems ability to recreate the motion environment close enough due to the shock & vibration experienced whilst operating is higher sea states. This limitation not only effects the environment but also learning skills such as wave / throttle technique. This inability to accurately simulate the motion environment has the potential to result in a negative effect of training.

4.2 (g) Virtual Reality

Virtual reality systems are becoming more readily available as hardware and software costs reduce. The benefits of being able to use a small portable system that provides a 360° view of the environment opens up numerous training possibilities, but brings with it its own interaction issues with virtual controls and real world controls. However this is being gradually addressed, e.g. the increasing use of mixed reality systems, bringing new possibilities to HSC simulation.

4.3 ON-WATER

On-water training will always be seen as the first and foremost option to train high-speed craft operators. Historically this was not an issue, but as craft have increased in speed and capability, the burden on the operators has increased to extreme levels. This increase is not just in speed, but also crew workload, as small craft now carry complex information systems whilst having to cope with reduced decision-making times.

5. STANDARD OPERATING PROCEDURES

5.1 GENERAL SOPs

The ability to practice Standard Operating Procedures (SOPs) prior to operations and in increasing demanding situations is vital to develop and maintain the required operational capability.

5.2 DYNAMIC NAVIGATION

A navigation methodology, termed DYNAmic NAVigation (DYNAV), has been developed to support high- tempo littoral operations [6, 7]. The navigation pair is not essential for the methodology to work, it is still useful if applied of a single crew. Evidence has demonstrated that as speed increases the operation of a HSC by a single person becomes ineffective, and thus high risk, as they are unable to undertake the required command and control tasks concurrently. The four phases of the DYNAV methodology are shown below and illustrated in Figure 3:

1. Plan (Navigator)
2. Communicate (Navigator to Driver)
3. Execute (Driver)
4. Control (Feedback between Driver and Navigator)

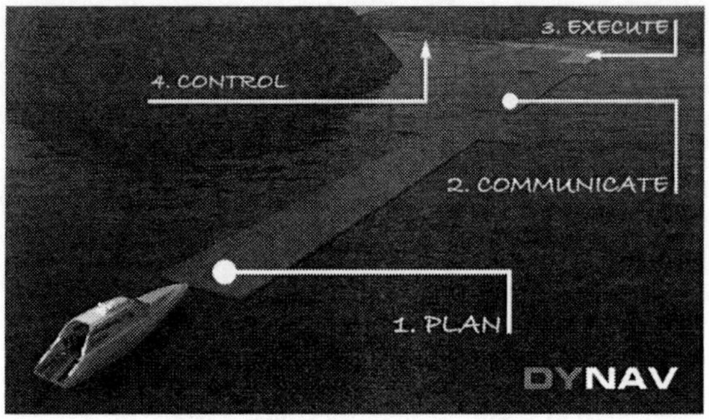

Figure 3: An Example of the Four Phases of the DYnamic NAVigation (DYNAV) Methodology.

DYNAV doesn't introduce any new navigational techniques. What it does is support the development of the competencies that make the difference between success and failure when the crew already has a level of technical competence. One perspective of vital importance is that DYNAV is not a technique that focuses on fixing standard errors or malfunctions. Instead DYNAV focuses on developing the system's (the integration of the craft and crew) total resilience.

The standard DYNAV communication protocol is a control strategy that supports the integration and assurance of the total system. In addition to the highlighted information in the protocol there are more subtle communication aspects / cues that support system resilience. This may be demonstrated by a competent driver reducing the craft's speed on their own initiative when the communication pattern alters from the normal / expected tempo. Such an alteration in verbal cadence can be a sign of an anomaly, or a problem, and the way the communication is conducted can be an early warning of such a situation even before the crew can definitively identify the problem.

Today, DYNAV training, for example within the Swedish Amphibious Corp, requires considerable time and resources. Within most professional organisaitons, on average, it will take a student at least 75 hours of dedicated, intense and focused navigation training (all other time excluded) to reach the required level of skill/competence to be considered competent / fit-for-role. To achieve this, a training time of between 12-16 weeks is typically required. But subsequent to this the crews require continual practice and coaching to maintain and enhance their proficiency.

6. INDIVIDUALS AND TEAMS

The essential aspect for the development of competence is the training curriculum that develops individual and team competence. HSC operations have been identified as being a two-person task above relatively slow speeds. This highlights the requirement for simulation to train the Navigator and Driver in their two different roles, at the same time, on the same system. However the individuals need to train for both roles, the Navigator and Driver, as they will swap roles at times during an operation.

7. CONTINUEING PROFESSIONAL DEVELOPMENT (CPD) & COACHING

As with many organisations who require the maintenance of competences in order to maintain standards of product delivery and safety, the implementation of Continued Professional Development (CPD) is a cornerstone of their quality control process. CPD not only supports the maintenance of current skill requirements, but also facilitates the learning of new skills, which could be vital to an organisations future. Simulation provides CPD with a tool which is able to train and practice specific skills in a repeatable and observable safe environment. This ability to review performance at a detailed level, analogous to professional sports coaches, can give small but vital increases in skill and performance. As highlighted above, CPD and simulation must address the specific KSA's which need to be practiced and rehearsed in order to maintain the required standard, and not just become the annual 'tick-box' exercise that CPD can unfortunately become.

8. ASSESSING COMPETENCE

The training instructors will assess the students at the end of their course and give a pass or fail assessment of their abilities. For this it is essential that the assessing organisation defines the metrics by which the pass / fail assessment is achieved.

It is essential that the training instils the crew members with confidence in their abilities. Those working in the SAR and naval sector will be working in poor sea conditions, at the edge of the craft / systems operating envelope and therefore they must have confidence in both themselves, their crew and there craft.

To assess the ability of a simulation system to enhance crew confidence, and subsequently competence, students were assessed before and after training during two different courses, one incorporating simulation (i.e. classroom, simulation, on-water) and the other only using classroom and on-water education and training methods. The assessment used two similar 45-minute realistic SAR Scenarios where the crew of three had to perform the following steps:

1. Listen to a *Mayday* broadcast and record the key details in the message.
2. Communicate with Coast Guard Radio and receive their tasking from Rescue Centre.
3. Plan their route to the scene on the paper chart and transfer that route to the plotter
4. Conduct a pre-departure inspection of vessel systems and personal safety equipment.

5. Depart and communicate an ETA to scene.
6. Navigate through a fleet of fishing vessels and nets
7. Transit across a straight in low visibility with opposing shipping traffic
8. Identify risk of collision using radar
9. Manoeuvre according to rules, to avoid collision
10. Arrive on rescue scene and assesse the scene
11. Conduct a person recovery approach
12. Debrief mission

The simulator used was a fixed-base simulator (Virtual Marine Technology, Canada) that replicated the crew workstations and layout of the SAR RIB used by the crew. The simulator system is shown in Figure 4.

Figure 4: The fixed-based simulator located at the C-CG RHIOT School and used for the training assessment.

The assessment [8] showed that the students gained a higher level of confidence when simulation was incorporated into the training programme. The improvements in the task related functions gained by the simulation condition, compared to the no-simulation condition, are illustrated in Figure 5.

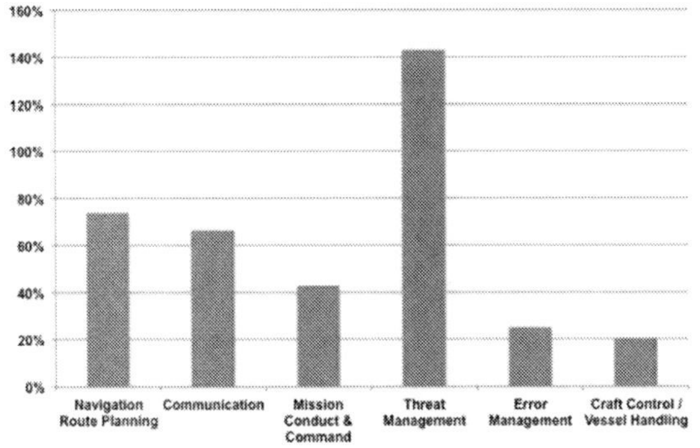

Figure 5: The difference in individual confidence scores illustrating the enhancement provided by the use of Simulation Within the Training Programme.

The results support the development of crew KSAs / functions in the following ways:

- Threat management (143%) – the crew can learn faster in a simulation environment where they can crash and learn from their mistakes
- Communication (67%) – they can repeatedly practice the standard DYNAV communication protocol
- Navigation route planning (74%) – The simulation allowed crews to practice the route planning skills learned in the classroom. Here knowledge of chart work and electronic navigation systems is key. The skills involved in chart work and keying in the route on the instruments are practiced and evaluated and finally the attitudes include diligence in checking for errors, taking the time to record routes information in the notebook as a backup for equipment failures.
- Mission conduct and command (43%) – they can practice working together
- Error management (25%) – they can see the results of their and system errors and have a better understanding of how to manage them
- Craft control / vessel handling (20%) – this is principally and practical task so no enhancement was expected

The Designers of this simulation-training program choose to focus the simulation scenarios on the operational areas of highest risk. In SAR operations teams are at the highest risk when in transit to the rescue scene. Especially when operating in conditions of restricted visibility or at night. The KSAs involved in identifying threats to the craft from collision with rocks or other vessels are best addressed by partial and full mission simulation. This is an area where the instructors cannot consistently place students into high risk situations in live training and therefore the simulator offers a consistent and repeatable method of exposing the trainees to situations they may experience on real operations. Due to the simulation focus on threat management it is not surprising that this competency is highlighted in the confidence results. It is essential to recognise that simulation must be properly incorporated into the training programme, with a specific curriculum and instructors capable and familiar with simulation training and it's seamless integration with on-water training.

It should be noted that the data is collected from real training and not a controlled scientific study. This situation does not allow for the use of a control group or allow for specific manipulations in examining different aspects of training, but does demonstrate the effective use of simulation within a real training programme.

9. CONCLUSIONS

Crew training is an essential part of delivering the required operational capability. Maintaining and increasing capability requires both system hardware and training to develop in unison and not one in advance of the other. Traditionally classroom education and on-water training has traditionally delivered the required competence but increasing craft speeds, operational complexity and tempo is demanding more skilled and capable individuals and teams.

Fully understanding HSC KSA's is required in order to correctly maintain and improve effectiveness in future operations. The correct implementation and integration of simulation and CPD in the HSC market is vital. Previously HSC simulation had not addressed the required KSA's or syllabus, therefore leading to systems based solely on familiarisation with new embarked hardware and not on the operation / task of the HSC.

The development and adoption of the DYNAV methodology has made it possible to implement a process for HSC crews which can be trained on simple and cost effective simulators, and not tied to expensive motion based systems. The ability to conduct CPD, both on-shore and on-water, now gives organisations the ability to conduct continual training without long gaps due to the availability or absence of previously expensive training systems. Simulation is now being effectively integrated with classroom learning and on-water training but it requires it's own specific curriculum and learning objectives that the instructors need to be engaged with.

10. REFERENCES

1. Hill, J., Dobbins, T., Myers, S. (2009) Advanced Coxswain Training. Conference Proceedings; Royal Institution of Naval Architects, SURV 7 - Surveillance of Search & Rescue Craft, 27-28 May 2009, Poole, UK

2. Hill, J., Forsman, F., Brand, T. and Dobbins, T. (2014) Risk, Competence, Interoperability and Qualifications For Fast Craft Operations'. Proceedings; RINA Human Factors Conference. London, UK.

3. Hill, J. and Dobbins, T. (2010) The Future Training of HSC Crew: Integrating Simulation and At-Sea Training. Conference Proceedings; Human Performance At Sea Conference, Glasgow.

4. Forsman, F., Dobbins, T., Hill, J., Brand, T. Dahlman, J. & MacKinnon (2012) Exploiting Simulation to Enhance Operational effectiveness and Interoperability in Littoral Maritime Operations. Proceedings: NATO CSO Symposium, Transforming Defence through Modelling and Simulation – Opportunities and Challenges" (MSG-094), Stockholm, Sweden.

5. Parker, R. (2007) We are in no doubt. USCG Proceedings. Fall 2007, pp 6-9.

6. Forsman, F., Dahlman, J. and Dobbins, T. (2011) Developing a Standard Methodology For Dynamic Navigation in the Littoral Environment. Proceedings; RINA Human Factors in Ship Design Conference, London, November, 2011.

7. DYNAV Manual. ISBN 978-1-5262-0347-2. www.dynav.org

8. Brand, T (2015) Unpublished data.

11. AUTHORS BIOGRAPHY

Jon Hill holds the current position of Director at Trident Marine Ltd. He retired from the Royal Marines as the Warrant Officer responsible for Landing Craft and high-speed boat training whilst also being responsible for craft and operational development. He currently works internationally providing training support to Governments/Coast Guards and advising on craft and training development.

Tyler Brand holds the current position of Director at Strongwake Ltd. He presently serves as a Canadian Coast Guard Officer, including responsibility for fast craft training. He subsequently worked developing innovative training solutions including the use of simulation integrated with classroom and on-water training.

Trevor Dobbins holds the current position of Director at STResearch Ltd. He has published widely on HSC human factors and is the principal author of the High Speed Craft Human Factors Engineering Design Guide. His previous experiences include projects for the UK MOD, RNLI, dstl, and the US Navy.

FROM REACTIVE IN TRAINING TO PROACTIVE IN DESIGN: APPLYING STANDARD MARITIME DESIGN

M Lutzhoft, University of Tasmania, Australian Maritime College, Australia, **M R Grech**, University of Tasmania, Australian Maritime College, Australia, **M Jung**, Korean Institute of Maritime and Fisheries Technology, Korea

SUMMARY

As advanced technology is increasingly being implemented on ships, variability and complexity in design and use also increases. Traditionally, so does training requirements. With sometimes more than 30 manufacturers on just one ship's bridge this variability is evident both within and between ships. Such variations have significant impact for crew, creating the need for more mandatory training requirements, negative transfer of learning and increased error rates, overall hindering effective decision making and degrading safety of navigation. Maritime accidents highlight the consequences of lack of standardisation and human centred design in navigation systems. The time has come to suggest that we cannot train away the problem, but should apply a more effective countermeasure to failure – design the problem out. The standardisation of navigational systems is one effective countermeasure to reducing variability and system complexity with the IMO now proposing guidelines on standardisation (S-mode guidance) as one of its outputs for 2018. Work on developing S-mode guidance within the IMO is now underway. This paper will provide an overview of research and development activities being undertaken to support the formulation of the S-mode guidance.

NOMENCLATURE

AIS	Automatic Identification System
CPA	Closest Point of Approach
EBL	Electronic Bearing Line
ECDIS	Electronic Chart Display & Information System
ENC	Electronic Navigational Chart
GNSS	Global Navigation Satellite System
HCD	Human Centred Design
INS	Integrated Navigation System
TCPA	Time to CPA
VRM	Variable Range Marker

1. INTRODUCTION

Recent statistics published by IUMI [1] indicate that the number of ships deemed *total loss* (a "write-off" where the lost value of a ship exceeds the value of the insurance policy) was at an all-time low in 2014, but increased during 2015. The causes which show a marked increase are heavy weather and grounding. Grounding is in its simplest form due to loss of awareness of the ship's position (or occasionally loss of propulsion). To aid a navigator's awareness, a range of equipment, more or less automated and/or "intelligent" is introduced to ship control rooms and bridges.

The number of manufacturers of maritime electronics equipment has increased substantially in the last decade. Without proper guidance, this has invariably led to complex and sophisticated technical systems in bridge and engine control room equipment design with significant variation in information displays, controls and system functionality. Since the instrumentation and configuration of information differs between ships, the probability of errors in understanding the presented information is high and can occur for both officers and pilots. This issue is further compounded by a poor status/mode feedback loop in current systems.

As a comparison, one seldom finds those problems in aviation because instruments and equipment are standardised. A pilot can move between different types of airplanes and will find that cockpit information and interaction is basically similar. Lack of understanding of complex integrated systems, what Woods calls buggy knowledge [2], has been a factor in several maritime accidents. Examples include the *Royal Majesty* (1995), *CFL Performer* (2008), *Seastreak Wall Street* (2013), *Ovit* (2013).

This issue has been highlighted by the Nautical Institute for some time. Already in the early 2000's it was found that end-users of bridge instruments considered interaction with new technology demanding. This can be due to varying designs emanating from different manufacturers, and research shows that seafarers have to work hard to make some sense of the instruments that pertain to support their work [3].

Neither regulation nor education can keep up with innovation. Education in system handling and system safety is getting difficult for educational institutions to handle. In most maritime academies there is one or perhaps two manufacturers' brand/s used for simulation training. Since different brands have different interaction paradigms, the education is always underspecified. For some of the most specialised equipment 10-day training courses are mandatory – only to learn their operation. This is over and above 3-4 years in a maritime academy and additional months at sea to become a fully qualified navigating officer.

The standardisation of navigational systems is now being encouraged as an effective countermeasure to reducing variability and system complexity. The introduction of a *standard mode* (S-mode) will simplify the use of complex maritime technology – making for less errors and will save time and money. It will also standardise training around the world creating a more effective mariner workforce.

1.1 EARLY DAYS

An initial proposal submitted to the IMO in 2008 by the International Federation of Shipmasters' Associations (NAV 54/13/1-The concept of S-Mode for onboard navigation displays) was to develop S-mode as a default standard display which would activate and present a standard user interface (e.g. north-up display, relative motion and true vectors). The initial idea was that when you pressed an "S button" the display including menus and information presentation would change to an across-brands standard. As learning progressed, more layers or controls could be added.

In the first known published study, 54 seafarers participated in a survey that was used to query needs for basic or default settings and controls [4]. Data was collected from seafarers, regarding design details on separate pieces of equipment. Results showed a relatively high variation in responses, which may be due to the open question format.

This project expands on earlier studies by using human-centred design (HCD), collecting a broad sample of multi-national user needs, making and testing prototypes and integrating results into a guidance document for submission to the IMO and for use by equipment manufacturers.

1.2 S-MODE DEFINITION DEVELOPS

The concept of S-mode as a default setting for navigation displays was proposed under the agenda of the IMO e-Navigation initiative. In June 2015 at MSC 95, IMO decided that the 'Guidelines on standardized modes of operation (S-mode)' should be given priority and instructed for this to be established by 2019. Further, a paper submitted to the IMO by a consortia of interested parties (document NCSR 3/28/1) on the development of guidance on the Standardized (or S) Mode of operation of navigation equipment provided the scope and direction under which the guidelines should be developed.

Recent workshops have assisted in refining the scope of S-mode with concerns raised over the rationale of a single button approach. The consensus was that "less emphasis should be placed on the activation of 'a single button' - instead focus should be on guidance that will better achieve standardization of the 'look and feel' of navigation systems". This resulted in the following definition being proposed: "*Standardisation of design for navigation and communication systems, encompassing displays, interfaces, and functionalities able to provide the bridge team and the pilot with timely access to essential information for the conduct of navigation throughout the voyage, from berth to berth.*"It is recognised that there is not a single optimal solution for any user interface. Even by following good design guides that optimise usability still means that quite different solutions will emerge. However, it has been agreed that a major role of S-Mode is to achieve more commonality in user interfaces to ease the ever-growing complexity of bridge familiarisation and that of operation. Now the consensus is to work toward a standard design, be it a process or the outcome of the process.

S-mode is now seen as a design concept for shipboard equipment that takes into consideration human centred design principles ensuring consistency and intuitive design, expediting (effective and efficient) familiarisation (training, continuous learning, elimination of type specific training) and supporting situation assessment and decision making.

It is apparent that more training, procedures, warnings and audits are not addressing the issue of error management and costly accidents are on the rise [5]. Hence it may be time to adopt a systemic view, so we can help people manage and make fewer errors. With layout, instrumentation and configuration of information vastly different between ships bridges and engine control rooms (observations by the first author identified that the number of brands on a bridge can be up to 30), errors in understanding occur for users, especially those who have recently joined a ship, or marine pilots who "join" new ships frequently – several times a day; thus they may not be able to handle situations that require take-over on short notice. The situation is similar in the engine control room in which systems have been designed with little consideration of the human factors with potentially similar consequences [6].

2. METHOD

This project expands on earlier studies by using human-centred design (HCD) as a methodology, collecting a broad sample of multi-national user needs, making and testing prototypes and integrating results into a guidance document for submission to IMO and for use by equipment manufacturers. HCD is a cross-disciplinary approach which involves users throughout the design process. It is important to note that HCD is not about asking end users to answer design questions, but is context based; information is gathered to describe their work and their tasks in their operating environment so that researchers and engineers can design possible solutions to be iteratively developed and evaluated. Data will be collected from cadets and active seafarers, both deck and engine crew.

2.1 PROCEDURE

In order to understand and specify users' needs on S-Mode, a survey was conducted on navigational equipment targeting maritime users between October and November 2015. This research is considered part of the design process through the use of HCD by collecting users' requirements for standardising navigational systems.

The questionnaire aimed to evaluate the participant's ratings on the need for standardisation of navigational display and each main task of the Integrated Navigational Systems (INS). The functions of "Route Monitoring" and "Collision Avoidance" were selected as a start, based on the IMO's Performance Standards for INS [7] as shown in table 1, Part 3 below. "Route Monitoring" refers to the navigational task of continuous monitoring of own ships position in relation to the pre-planned route and the water. "Collision Avoidance" includes tasks such as detecting or plotting other ships or objects to avoid collision. There are other functions related to route monitoring tasks of INS, however the survey items on the questionnaire were selected primarily based on the IMO's Performance Standard of INS.

Table 1: Main items of the questionnaire

	Questions
Part1	Rank and experience with navigational equipment
Part2	Masters and navigators' perception of the need of S-Mode (the scope of the S-Mode)
Part3	Subjective ratings for standardising detailed tasks of INS • Route Monitoring (MSC.232(82), IMO require ments on ECDIS) • Collision Avoidance (MSC.192 (79), IMO req uirements on RADAR)
Part4	Designing the display of navigational equipment

Reponses were received from fifty Korean seafarers with sufficient on board experience.

When completing the questionnaire, respondents could reply by selecting their preference of setting functions of each main task of INS. Related to the design of the navigational display, the respondents were also asked to draw on an ECDIS mock up screen on the questionnaire. Three elements were nominated; Menu Bar, Route Information and Alarm Information.

3. RESULTS

3.1 STATUS OF RESPONDENTS

The fifty respondents comprised of maritime pilots, masters and deck officers as shown in Tables 2 and 3. All respondents had seagoing work experience with seventy-two percent of respondents having ten years or more of seagoing experience.

Table 2: Rank of respondents

Rank	Respondents	Proportion
Maritime Pilot	9	18.0 %
Master	14	28.0 %
Deck Officer	24	48.0 %
Shore	3	6.0 %
Total	50	100.0 %

Table 3: Years of sea-going service

Year of experience	Respondents	Proportion
1 ~ 4 year	4	8.0 %
5 ~ 9 year	10	20.0 %
10~14 year	11	22.0 %
15~19 year	3	6.0%
20~24 year	9	18.0%
25 ~ 29 year	4	8.0%
> 30 year	9	18.0%
Total	50	100.0 %

3.2 USER REQUIREMENTS ON THE NECESSITY OF S-MODE

Regarding the necessity of implementing S-Mode, the respondents rated the need for each navigational aid, i.e. 1 is the least necessary and 5 is the most necessary. The survey findings shows that the bridge instruments selected as most in need of standardisation included Electronic Chart Display and Information System (ECDIS), INS and Radio Detection and Range (Radar), closely followed by Automatic Identification System (AIS), Global Navigation Satellite System (GNSS), Echo sounder and Speed log (Figure 1). The results indicate that seafarers regard ECDIS, INS and RADAR as having a higher priority than other systems to be standardized.

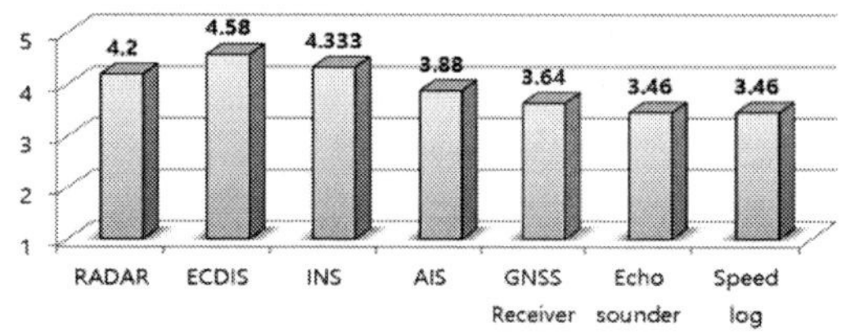

Figure 1: Findings of participant's ratings on standardizing navigational equipment

3.3 USER REQUIREMENTS FOR NAVIGATIONAL TASKS OF INS

Part of the survey requested feedback on the prioritisation of functions for the tasks of route monitoring and collision avoidance that should be standardized.

The tasks of Route Monitoring were evaluated by users by rating the necessity of standardization of each tasks. Table 4, shows the mean rating given by respondents (1 is the lowest needs rating and 5 is the highest needs rating). The results convey the S-Mode needs rating applied to all of the fourteen tasks. Moreover, the following tasks were given a higher rating (4.5 or above); as the most preferred to be standardized for Route Monitoring.

- Alarm setting for passage of safety contour
- Setting of three categories for ENC display - base, standard, and others.
- Setting of a symbol of own ship (Vector)
- For route monitoring, the feature to return to the previous display where own ship is indicated at the display's center
- Calculation methodology of direction and distance between two points (EBL, VRM)
- Functions to update ENC
- Radar, ARPA and AIS overlay

Table 4: Evaluation of Route Monitoring tasks

Tasks (Least needed (1) – Always needed (5)	Mean
Setting of three categories for ENC display - base, standard, and others.	4.5
Setting of safety contour	4.66
Setting of scale adjust	4.44
Setting of a symbol of own ship (Vector)	4.52
For route monitoring, a feature to return to a display where own ship is indicated at the center	4.673
User chart editing – dots, lines, faces, texts, symbols	4.4
DR setting.	3.82
Manual calculation of LOP (Line Of Position)	4.24
Voyage data recording.(Time label)	4.26
Calculation methodology of direction and distance between two points (EBL, VRM)	4.52
Functions to update ENC.	4.673
Radar, ARPA and AIS overlay.	4.633
For route monitoring, alarm setting	4.32
Selecting an alternative route	4.085

The setting of functions for collision avoidance was also rated by respondents as shown in table 5. The study shows that CPA (100%) and TCPA (90%) alarm setting were given the most preferences followed with trail (86%) and off-centering (84%). In contrary, setting range ring was set as 6 miles by 56 percent of respondents. The result gives an understanding on user preferences when they perform tasks of collision avoidance.

Table 5: Evaluation of Collision avoidance tasks

Function	Result
Range Scale	56% of respondents preferred 6 miles
Gain	60% of respondents preferred auto configuration
Clutter	58% of respondents preferred auto configuration
Motion of center	64% of respondents preferred True Motion
Off-centering	84% of respondents preferred the application of Off-Centering
Trail	86% of respondents wished to have it indicated, and preferred both 3-minute and 6-minute time set
Vector	70% of respondents chose True Vector, and 56% of them preferred 6-minute time set
Display mode	Almost 75% of respondents wished to set Display Mode in N/up mode
Band	80% of respondents wished to set Band in X-band mode
CPA alarm Setting	100% of respondents stated that CPA Alarm has to be indicated and 30% of them wished to set the distance in a mile measurement
TCPA alarm Setting	90% of respondents wished to have an indication, and most of them preferred 10-minute time set

3.4 DESIGNING NAVIGATIONAL SCREEN

To design the ECDIS screen, a blank ECDIS mock up screen was included in the questionnaire. Respondents were asked to draw given elements such as Menu Bar, Alarm Information and Route Information. For the analysis, the screen was divided into the nine parts. The result is shown in figure 2. The study shows where the respondents prefer to position each element on the screen.

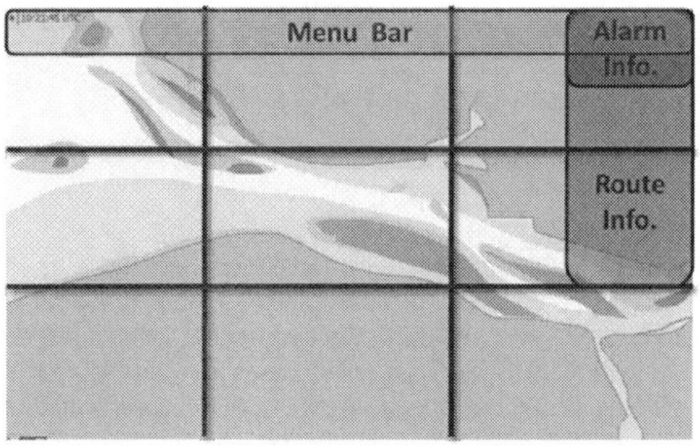

Figure 2: The result of designing an ECDIS screen

User preferences on settings of display during Route Monitoring were surveyed as shown in table 6 below. The priority in the table were selected by respondents. The heading information and the speed over ground were given the highest priorities among the tasks. On the other hand, setting range ring is shown as the lowest mark and following lower tasks are the rate of turn in number and the route name for route monitoring. This result can be utilized to design layout of the information of a Route Monitoring screen. The information having higher priorities can be displayed in more optimum positions.

Table 6: Preferences of Route Monitoring display settings

Function	Yes	No
Heading information in number	50	0
COG (Course over ground) in number	49	1
SOG (Speed over ground) in number	50	0
STW (Speed through water) in number	41	8
UKC (Under keel clearance) in number	43	7
ROT (rate of turn) in number	39	11
The route last selected	41	9
Route name for route monitoring	38	11
Course, direction, distance, and time to waypoint	46	4
Distance and time to WOP (wheel-over point)	46	4
ETA to the last waypoint (estimated time of arrival	48	2
Total length of route	42	8
Course info after target waypoint change	47	3
Range ring	34	16
Voyage recording	40	10
AIS information on Aid to Navigation	43	7

4. DISCUSSION

Challenges include the understanding of what standardisation means to the maritime community, where manufacturers wish to avoid it for sales of equipment and type-specific training purposes as well as avoiding lock-in of innovation but it is now clear that seafarers wish both for individualisation and standardisation of navigation systems. Complete consensus however on the meaning of S-mode is under discussion. Some insist that for S-mode to succeed the 'look and feel' aspects need to be fairly defined with some level of prescribed standards necessary. However, a number of Maritime Administrations have moved away from a requirement for 'Type Specific' training in place of 'demonstrated familiarisation' (as yet undefined). The support for utilising the Human Centred Design process as a guiding principle for defining S-mode is undisputed. What is still being discussed is the level of 'standard' that is required to achieve the original intent of 'standardisation'. The final outcome could possibly well end up being a combination of a 'process' and a 'product'. However whether this will have the desired effect is still up for discussion.

5. CONCLUSIONS

The introduction of a "standard mode" is intended to simplify the use of complex maritime technology – making for less errors and time and money saved as well as more effective training due to a standard system and instructor expertise. S-mode will allow for a common training program in maritime training centres, easier familiarisation on board for new officers, quicker joint awareness for teams and speedier inclusion of a pilot into the bridge team. It also sets a precedent for future complex systems in all areas of a ship.

To gain a more complete picture we need to study what the work on the ship is today and may be expected to be in the future. We believe that using prototypes based on contextual scenarios in future development may guide the users closer to a "consensus" design.

This research is aimed at specifying three major parts, which are formulating the necessity of S-Mode on bridge instruments, the users' requirements of detailed navigational tasks of INS and the design of navigational screen to be considered in future development of S-Mode. Parallel work on engine control room design is also being undertaken to ensure a balance is achieved in the application of human centred design in other ship control, systems.

6. REFERENCES

1. IUMI Casualty and World Fleet Statistics Analysis. http://www.iumi.com/ 2016

2. Woods, D. D., Johannesen, L. J., Cook, R. I., & Sarter, N. B., *Behind Human Error: Cognitive Systems, Computers, and Hindsight*. CSERIAC SOAR 94-01. Ohio, USA: Wright-Patterson Air Force Base. 1994

3. Lützhöft, M., "The technology is great when it works": Maritime Technology and Human Integration on the Ship's Bridge, Linköping University: Linköping, 2004. Available at: http://www.diva-portal.org.

4. Jacobson, E. and Lützhöft, M., Developing User Needs for S-Mode, NAV08-ILA37 International Navigation Conference, London, 28-30 October, 2008.

5. Abeysiriwardhane, A., Lutzhoft, M., and Enshaei, H. Human factors for ship design; exploring the bottom rung. *International Journal of Marine Design, 156*(C1), 153-159, 2014

6. Mallam, S, and Lundh, M. The physical work environment and end-user requirements: Investigating marine engineering officers operational demands and ship design. Work (Preprint) 2016.1

7. IMO, Adoption of the revised performance standards for integrated navigational systems (INS), Resolution MSC.252(83), 2007

7. AUTHORS BIOGRAPHY

Margareta Lützhöft is a master mariner, trained at Kalmar Maritime Academy in Sweden. After leaving the sea, she studied for a Bachelor's degree in Cognitive science and a Master's in Computer Science. In December 2004 she received a PhD in Human-Machine Interaction. Presently she is holding a position as Professor of Nautical studies at the Australian Maritime College. Her research interests include human-centered design and the effects of new technology.

Michelle Grech holds an adjunct position as Associate Professor at the University of Tasmania, Australian Maritime College. Michelle also Heads the Human Factors Section at the Australian Maritime Safety Authority. She is responsible for leading the integration of human factors in shipping, currently supporting the development of the S-mode guidelines for e-navigation. Her previous experience includes managing Human Systems Integration research for naval platforms, maritime surveyor and shipyard commissioning engineer.

Min Jung is an assistant professor in Korea Institute of Maritime and Fisheries Technology since 2011. She is responsible for teaching seafarers in navigational knowledge and researching on S-Mode. Dr. Jung received Ph.D. of Maritime Traffic Information Technology in Korea Maritime University in 2008. She sailed international voyages as a deck officer in LPG tankers and training ships (2001-2005) and worked as a surveyor in Korea Register of Shipping (2008-2011).

HUMAN FACTORS DIMENSIONS IN THE DESIGN OF A PCTC VESSEL

N Bialystocki, Antares Shipping Ltd., Israel

SUMMARY

Based on the results of a paper by Costa and Lützhöft (2014), The Values of Ergonomics in Ship Design and Operation, presented at the RINA International Conference on Human Factors in Ship Design & Operation in 2014, this paper exemplifies the practical consideration of Human Factors and Ergonomics (HF&E) Dimensions of Ship Design using real-world illustrations from the design of a Pure Car and Truck Carrier (PCTC) vessel.

Some illustrations included in this paper stem from feedbacks submitted by the crew considered during detailed design stage, and some go beyond these, based on the design team's focus on the HF&E dimensions during the entire de sign stage, and mostly during detailed design stage.

The perceived benefits of considering HF&E dimensions of ship design are hereby experimented and substantiated using the case study of a PCTC vessel. It is believed that demonstrating real-world Human Factors implications in the design of a vessel is essential for promoting the importance of Human Factors and the possibilities that having Human Factors as a mind-set can produce in ship design.

NOMENCLATURE

HF&E	Human Factors and Ergonomics
ILO	International Labour Organization
IMACS	Integrated Monitoring Alarm and Control System
IMO	International Maritime Organization
ISO	International Organization for Standardization
MLC	Maritime Labour Convention
PCTC	Pure Car and Truck Carrier
SOLAS	International Convention for the Safety of Life at Sea
TS	Technical Specifications

1. INTRODUCTION

Based on results by Costa and Lützhöft [1], a set of benefit categories and dimensions of ship design are recognized to support the naval architect towards an ergonomic design. These dimensions represent the abilities of the ship to answer the needs of those who operate it: *workability, habitability, maintainability, survivability, controllability* and *affordability* (see for definitions). More so, HF&E dimensions of *communication and user empowerment*, and *professional and psychosocial aspects*, are captured [1].

Certain aspects interpreted to affect *workability* are also

Table 1: Definitions of the HF&E dimensions of ship design [2, 3, 4, 5].

Workability	Conditions onboard that support the seafarers to perform their tasks, including equipment (hardware and software), materials and procedures, physical and social environments, information, handbooks, and language.
Habitability	Adequate, comfortable and practical accommodation, cooking and washing facilities, storage and recreational spaces having regards for size, shape, gender, culture and environmental stressors such as noise, temperature and vibration
Maintainability	Conditions onboard that allow seafarers to perform the necessary maintenance of the ship, including access, tools, through-life support for the lifespan of the ship, and the design of operational maintenance tasks to be safe and efficient
Survivability	Availability of adequate equipment and facilities for firefighting, damage control and lifesaving, and the capabilities of the crew to ensure safety of crew and passengers
Controllability	Refers to the integration of users with equipment and interfaces, and appropriating layout of work stations, communication facilities, controls, displays, alarms, lights, etc. to allow the seafarer to perceive the status of machines and systems and provide fitting responses
Affordability	Refers to the total ownership costs associated with system/technology redesign, manpower, training, human support, and reduction of errors and accidents

focused and discussed upon below, having been considered during the design of a Pure Car and Truck Carrier (PCTC) case study vessel: *physical ergonomics and usability; efficiency; jobs, tasks, procedures and routines; harmonization and standardization; variability and flexibility.*

This paper describes the real-life design case of a new PCTC vessel having implemented and assessed the Human Factors and Ergonomics (HF&E) dimensions of ship design. This case represents a success story of HF&E integration and may serve as an example to other maritime organizations aspiring to expand their human-centred design implementations.

2. THE PCTC VESSEL DESIGN CASE

The ship owner designer, Antares Shipping Ltd. is a consultant company involved mostly in the planning, construction and operation of PCTC's fleet. Once the client decided to build a new PCTC vessel, Antares Shipping Ltd. developed a new concept design, in collaboration with the client, the shipyard, and other ship designers. The difference in the PCTC design compared to older designs is mostly attributed to:

1. Panama Canal Expansion Project, which allows vessel's breadth to be widened up to 49.0 meters.
2. Market demand for more efficient PCTC vessels, i.e. having less DFOC (Daily Fuel Oil Consumption) per Car Equivalent Unit (CEU).

The PCTC vessel was built in South Korea, and this was done after several design iterations carried out for having the most efficient PCTC vessel in the market. At last, the principal characteristics and main specifications of the PCTC were settled as follows (see
Figure 1):

Length overall:	199.95 meters
Breadth:	35.40 meters
Draft (Design):	9.00 meters
Draft (Scantling):	10.60 meters
Deadweight (Design):	11,766 metric tons
Deadweight (Scantling):	20,941 metric tons

No. of car decks:	13 (4 liftable)
Car deck area:	62,550 m^2
Car Capacity:	7,712 CEU
Service speed:	about 20.0 knots
Cruising range:	about 29,000 nautical miles

Complement:	32 + 6 persons
Flag:	Bahamas

As part of the new PCTC vessel design, various changes related to the Human Factors ship design dimensions and influencing aspects [1] were considered, having gained experience from previous PCTC designs of projects, and from the operation of a PCTC fleet.

Figure 1: General view of the PCTC vessel captured by a drone during sea trial in September 2014.

3. METHODS

3.1 THE DESIGN STAGES

The design process is generally divided into three main stages, chronologically ordered as follows:

a) Concept Design
b) Preliminary Design
c) Detailed Design

Antares Shipping, as the ship-owner's representatives, led all three design stages. Usually, outline specifications are signed by the ship-owner and the builder, between 1st to 2nd design stages, i.e. between Concept and Preliminary designs. Furthermore, the full TS of the vessel are signed by the owner and the builder, between the 2nd to 3rd design stages, i.e. between Preliminary and Detailed designs.

Moreover, the detailed design stage normally also expands after the vessel's steel cutting and start of construction. It is normal that building of a new vessel is started before completion of the detailed design. In reality, design changes can be made until the delivery of the vessel. However, in later construction stages, the feasible modifications are more and more limited, although when a series of vessels is built, modifications can be applied to subsequent ships.

In the case of this PCTC design, the main events of construction were as follows:

a) Steel cutting – November 2013.
b) Keel Laying – March 2014.
c) Launching – June 2014.
d) Delivery – October 2014.

The detailed design sent for owner's comments and approval consisted of 312 drawings of the following subjects:

a) Basic plans
b) Hull Plans
c) Hull Outfitting (including cargo handling equipment) plans

d) Accommodation plans
e) Hull Piping plans
f) Machinery plans
g) Electrical and Automation plans

At the time of steel cutting, about 54% of the drawings submitted to the owner were approved. Hence, it can be understood that the design continues during construction of the vessel as far as practicable, not to delay the construction schedule.

Every modification in the PCTC vessel design that was agreed upon between the shipyard and the owner was recorded, and during the detailed design stage a total of 1,716 design modifications were applied, each of these related to the HF&E dimensions of ship design.

3.2 USER FEEDBACK

Crew feedback was gathered from their experience at the ship owning company, from running other PCTC vessels.

The feedback is mostly ideas of how to improve any or several of the HF&E dimensions on board. The crew is encouraged to pass their feedback either good or bad onto the ship's manager, and to the ship-owner, with their proposal for design improvement. In some cases the crew passes it directly onto the ship-owner's design team. Once a request for a design improvement is received, the ship-owner's design team will pass it onto the shipyard with (sometimes) a proposed solution. Once a solution is agreed upon between the owner and the shipyard, the loop is closed by informing the crew about it.

The collection of such crew feedback took place, mostly but not only, during detailed design stage, described above, which is the most prolonged design phase and enables the owner and the crew to come up with ideas to improve the future vessels based on accumulative experience. Design decisions during the first stage are mostly related to the PCTC's shipping market, for determining the vessel's main particulars, and therefore the involvement of crew is less relevant. At preliminary design stage, which is the second phase out of three, the outline specification will be determined, mostly based on the minimum rule requirements, with some improvements based on crew feedback (e.g. installation of elevator).

On the other hand, the detailed design is very much related to the specific operation of the PCTC vessel, and therefore inputs from the users were highly appreciated and seen with very much importance. Moreover, the detailed design stage is usually welcoming ideas and innovations which are on top of the market and rules requirements.

It is important to note that various rules and regulations exist in the shipping industry, and many of those are related to the human element. However, the regulations represent a minimum threshold, for the crew safety, efficient vessel operation and maintenance, comfort, etc. Throughout this paper, examples which exceed the regulations are provided, based on the ship owner designers' accumulated experience.

3.3 ASSESSING USER FEEDBACK IN TERMS OF HF&E DIMENSIONS OF SHIP DESIGN

As the ship-owner and shipyard were going through the crew feedbacks, the HF&E dimensions of ship design were utilized to help think of the ideas for improvement from a broader perspective of the ship's abilities to fulfil the needs of the crew and the specific operations it was intended for.

The design improvements analyses was carried out by the owner designers, in collaboration with the ship's manager and the crew, mostly by telephone and/or meeting discussions, where the solution was selected after brainstorming. Such solution was not always presented to the shipyard, in order not to dictate the shipyard how to solve a ship building problem. Some cases the yard came up with a similar solution to the one of the owners side, and in other cases, were the yard came up with a different solution, a further discussion was carried out between the owner designer and vessel manager and the crew, and so forth till the solution was accepted to all parties.

3.4 ASSESSING ADDITIONAL DESIGN IMPROVEMENTS IN TERMS OF HF&E DIMENSIONS OF SHIP DESIGN

Additional design improvements were made at the will of the design team based on their interpretations and perceived value of the HF&E dimensions of ship design. These additional design improvements sometimes encompass direct user input, and sometimes did not involve the crew directly, and this depends on the ship owners designers opinion whether there is a need for a direct participation of the crew.

Similar to design improvements raised by the crew, also in the case of improvements initiated by the designers, an analysis was carried out and a solution was lastly agreed between the owner and the shipyard. The crew was informed about such design improvements, and was welcomed to share from their thoughts and ideas.

The vessel's drawings' improvements applied to the case study PCTC vessel, were analysed in the light of the HF&E dimensions of ship design.

All design improvement, not matter what was their origin, were recorded in a list and spread to the crew for them to be aware of. As the crew is usually repatriated every 4 months, the list of improvement is spread to the ship manager for informing the next crews throughout the vessel's lifetime.

4. RESULTS AND DISCUSSION

The results are introduced here describing examples of artefacts of the PCTC vessel improved through the implementation of the HF&E dimensions of ship design during analysis of the received feedback from the crew members, along with other modifications that were made based on the ship owners design team's focus on the HF&E dimensions. The examples are illustrated, and how they were thought through and improved is explained. It is important to notice that the chosen examples are the ones that regular PCTC vessels are lacking of, with regards to similar aspects.

4.1 WORKABILITY

Below, practical design improvements compared to the running PCTC vessels are discussed with illustrations, with regards to the Workability dimension.

4.1 (a) Physical ergonomics, usability

It takes at least 30 minutes to ensure a good night vision of the navigational screens on a ship's bridge during night-time, and bright lights must be avoided during night-time watches [6] to ensure that the information on the screens can be visualized and safety maintained. The case PCTC bridge design includes dimmers to all lights and instruments in the bridge, as well as a thick double layered partition curtain between the conning stand and the chart and radio tables.

In order to improve the officer's visibility from the bridge, the forward deck seen from the bridge, was cleared from any obstruction and cargo hold fans were brought one deck below for allowing efficient look-out (see Figure 2).

Figure 2: View from bridge without obstructions.

For improving the usability of the bridge control console instruments, the design of the console was such that the hand can reach conveniently every control instrument, while the indicators are in the vertical part of the bridge control console, for being in front of the user's eyes. On the other hand, the wing control console was divided into two parts, namely: 1. Wing control console, which includes the Main Engine manoeuvring, the Steering Gear and thrusters' controllers. 2. Instruments console, having mainly indicators that are not needed to be hand reached. The control console was slim and attached to the ship's side, enabling the master/pilot to control the vessel while looking at the ship's side shell approaching the quay. The controllers were arranged logically in accordance with the orientation of the vessel.

For avoiding crew mistakes that may damage machineries, various innovative ideas were implemented in the design from early stages. An example is an interlocking that was prepared between the vessel's bow thruster and the bow thruster's room mechanical ventilation fan, and another interlocking between the fan and the fan damper. These were introduced in order to prevent starting the thruster while air is not circulated in the room, which might cause thruster motor trip, overheat, or even burn.

In terms of vision, reach and lift, CCTV was installed at critical locations in the vessel, as well as a stern ramp camera simulating the landing of opened ramp on the quay. This is highly important in order to prevent the ramp opening on quay side bollards. An elevator was added from the technical specification stage of the vessel, with many stops between the engine control room floor and the accommodation floor.

For improving the teamwork, the vessels was packed with telephones in every cabin and working room, much above nay rule requires. Public address speakers were added in each cargo deck, clocks all over the accommodation public spaces.

Weather and water tight doors and access hatches, were all designed with one operating turning wheel, instead of multiple dog clips. Moreover, the doors are all arranged with a slamming spring that enables to catch the heavy steel door before tightening it, in a way that the user doesn't need to press the door and turn the wheel simultaneously (see Figure 3).

Figure 3: Water tight door with slamming spring, and one wheel for easy handling.

In the engine room main transfer valves of water ballast and fuel are remotely controlled from the engine control room, in order to avoid crew running in the engine room to open/close valves.

The air conditioning system is considered critical system that enables the good operation of the vessel. For that reason the air handling unit is design with redundancy, i.e. two independent systems that condition the entire accommodation, and two units that condition the engine control room. The engine room workshop, which is normally attended during day shift, is also air conditioned.

Refrigerators and laundry equipment on board the subject PCTC vessel were arranged longitudinally, and this is in order to prevent their content to fall due to ship's rolling motion, which is stronger than pitching (see Figure 4).

Figure 4: Laundry equipment arranged longitudinally (note the all doors are opened to the same direction to avoid interference between doors).

4.1. (b) Efficiency

The vessel was designed to be as less complicated as possible, therefore un-used equipment and systems, which are normally provided by the shipyard, were removed. An example is the drinking water coolers (fountains), which are no longer used by the crew, because they prefer drinking water from bottles. At the same time, in order to prevent plastic bottles accumulation, a plastic granulator was provided for every vessel, otherwise the garbage room would be very soon packed with plastic bottles.

A main concept is that simple installation of systems will incur simple and efficient maintenance. For that reason, several units were designed in modules "plug and play" type such as fuel oil supply system and the fresh water unit. More so, the fresh water unit and fresh water generator were installed with a Hydropath Marine (HM) water treatment, which enables the crew to waive the chemicals cleaning cooler's plates in the sea water side of the fresh water evaporator, and waive the use of UV lamp in the fresh water sterilizer. Using of HM treatment saves maintenance time, and consumables on board.

The vessel's bridge is equipped with many instruments, each has its own alarm. With the intention of integrating systems, a Bridge Alarm Monitoring System (BAMS) was designed in order to concentrate all bridge alarms in one panel, that enables the officer on watch improve the situation awareness, while avoiding un-necessary movements in the bridge, which is extremely important mainly during emergency (see Figure 5).

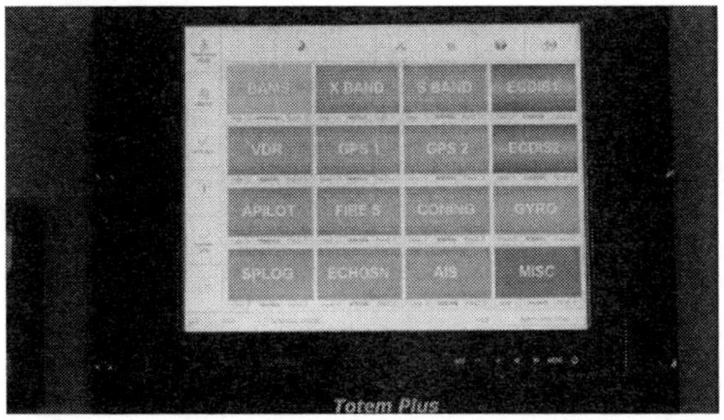

Figure 5: Bridge Alarm Monitoring System panel arranged in the bridge control console.

The vessel machineries are controlled from an IMACS, manufactured by Totem Plus. With this system the crew can easily monitor and control engines, aux. machineries (valves, pumps, etc.), cargo hold lights, cargo hold ventilation fan, tank's level, etc. The IMACS integrates the vessel's systems, making the work on board less complicated, having all information available in one system, which improves the operation of the vessel. The system has the option of setting of alarm set points, delay times and calibration of sensors by software. Totem Plus philosophy of using commercial Off-The-Shelf hardware eliminates the dependence on dedicated, more expensive and less accessible "marine" spare parts, enabling easy service and repair anywhere in the world [7].

4.1 (c) Job, tasks, procedures, routines

Table 2: The minimum and the designed number of crew on board

Grade	Flag Requirement	Yard's proposal	Owner's proposal
Master	1	1	1
Chief Mate	1	1	1
Nav. Off	2	3	3
A/B	1	5	5
Nav. Rating	3		
Cook	1	1	1
Chief Eng.	1	1	1
Second Eng.	1	1	1
Eng. Off.	1	2	2
Eng. Rating	2	5	5
Others	-	4	7
Total	14	24	27

The PCTC case vessel *Minimum Safe Manning Document*, in accordance with the flag safe manning requirements [8], is 14 crew members, however, in order to allow resting hours, an accommodation for 24 crew members was proposed by the yard, and finally a design for 27 crew members was accepted, as shown in Table 2.

All doors are always opened away from the cargo space, to prevent a person opening a door and hitting a car by mistake. Access hatches covers are also opened away from the cargo area.

All hull piping systems were designed passing in the ship's sides, and not disturbing the cargo loading or unloading. Where the cargo hold has a wall boundary, the fittings are recessed (see Figure 6), in order not to penetrate the cargo space and interfere with cargo flow in/out of the vessel. More so, sounding pipes were arranged passing inside deck supporting pillars, to protect the pipe and the cars (see Figure 7).

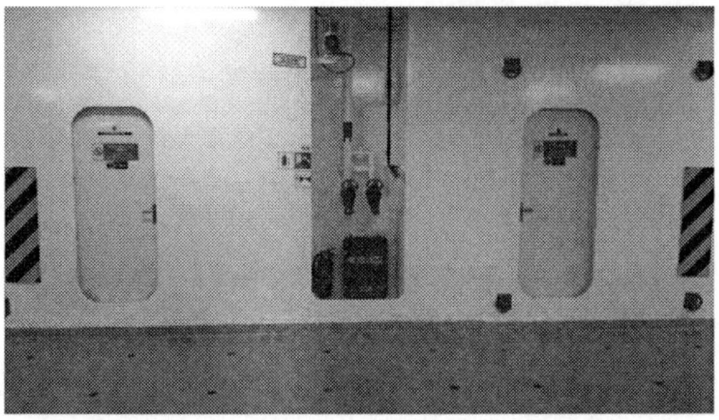

Figure 6: Doors in cargo space opened away from the cargo hold, and recessed fittings in cargo space to avoid cargo damages.

Figure 7: Sounding pipe led inside cargo hold pillar to protect the pipe and the cargo from interference.

Liftable decks are divided to panels, each panel is fixed on semi auto stoppers at 3 different heights. The stoppers are painted in 3 different colors, and the operating pulling ropes for the stoppers are with the same color of the corresponding stopper to prevent crew mistake of pulling the wrong rope.

Fixed ramps and corresponding deck openings inside cargo hold are arranged with kick plates on each side of the ramp to avoid fittings such as cargo lashing material rolling and falling to a bottom deck, with possible damage to the crew and/or to the cargo (see Figure 8).

Despite of various means to avoid cargo damage by human mistake, there are unavoidable structural members such as brackets that in some areas penetrate the cargo space. These are all marked with zebra sponge reflecting tape, to emphasize the danger.

Figure 8: Kick plates (yellow painted) around fixed ramp and around deck opening. Note also the yellow/black zebra markings that emphasize danger area to the stevedores.

The subject PCTC vessel has air draft limitation, due to under bridge passage when approaching Toyota car terminal in Nagoya, Japan. The limitation is 46.2 meters from the keel to the mast head top. The normal sailing position of mast is above the air draft limitation, and the crew needs to fold the mast before the entrance to Nagoya port. For having a more efficient mast folding procedure, it was decided to design and apply a hydraulically folded mast, with mechanical folding option as a back-up (see Figure 9). The crew is trained for this task, and it's a standard procedure that can reduce manpower.

Figure 9: Hydraulically foldable mast, before under bridge passage.

4.1 (d) Harmonization, standardization

The PCTC vessel was designed with the owner's standard makers list, because of the crew familiarity with the equipment manufacturers from previous vessels in the owner's fleet. The standard manufacturers that are incorporated in the design from the TS stage, supply the main equipment for the vessel's operation such as main engine, steering gear, bridge navigation and communication equipment, Control and automation system, cargo handling equipment, etc.

Standard equipment is used also for the maintenance of the vessel. An example is the grease nipples on board are all the same hexagon hook on type, which are also easily adapted to a grease gun. The crew has to grease all over the vessel regularly once a week, hundreds of points, and the use of standard grease nipple is a time saver, because there it doesn't require adapting the grease gun to various shapes of grease nipples (see Figure 10).

Figure 10: Standard grease nipple, hook-on type.

Another example of standardization is the Printed Circuit Boards (PCB's) of all equipment was decided to be from three manufacturers only. This enables easier maintenance by the crew, and saved managerial time in contacting various PCB manufacturers.

4.1 (e) Variability, flexibility

Adaptable equipment in way of multifunctional display is one of the categories within the workability HF dimensions. The vessel's IMACS is the best example for the flexibility, since there is one monitor showing about 22 mimic screens of various systems in the ship. The Electronic Chart Display and Information System (ECDIS) arranged in the wheelhouse, is interfaced with other navigational equipment such as: Radars, Gyro Compass, Speed Log, Echo Sounder, Differential Global Positioning System (DGPS), Automatic Identification System (AIS) and with the Anemometer (wind sensor). As such, the ECDIS monitor is capable of showing a vast variety of information simultaneously (see Figure 11).

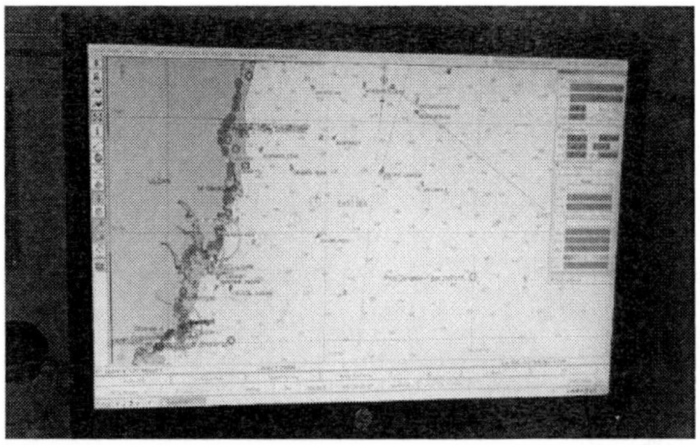

Figure 11: Electronic Chart Display and Information System control monitor, interconnected with other navigational equipment.

Most of the machineries and systems on board the PCTC vessel are controlled by software, and in many cases the parameters can be set in accordance to the user's personal preferences, as long as the minimum safety requirements are kept.

4.2 HABITABILITY

On top of the designed number of cabins for crew members, as shown in Table 2, two extra cabins were proposed by the yard, for Suez crew and for Pilot, however, five extra cabins were actually added to the design from the TS stage:

- 2 persons in owners cabin
- 1 person in Superintended cabin
- 1 person in Supercargo cabin
- 1 person in Pilot cabin
- 6 persons in Suez crew cabin.

The persons intended to be accommodated in these cabins are not normally on board. Nevertheless, they attend once in a while the vessel, and in order to have comfortable and user-friendly accommodation, dedicated private cabins, each with private toilet, were prepared.

The MLC 2006, which every vessel built today, must be in compliance with, state that electric light should be provided in seafarer accommodation and the standards should be fixed by the national authority [9]. The Bahamas Maritime Authority (BMA) on the other hand state that the standards of artificial lighting should be in accordance with the Merchant Shipping (Crew Accommodation) Regulations, 1978, as amended [10]. The last amendments came into force in 1997 and it can be seen, that in view of the ship-owners recognizing the crew comfort and social environment as high priority, the design of the subject PCTC vessel has illumination level above the minimum required by the flag (see Table 3).

Table 3: The minimum and the designed illumination level on board.

Cabin or space / Illumination level	Required by the flag	Proposed by the yard	Designed by the owner
Cabin	50 lux	100 lux	150 lux
Mess room	150 lux	150 lux	200 lux
Recreation room	50 lux	100 lux	200 lux
Hospital	50 lux	100 lux	100 lux
Offices	100 lux	200 lux	200 lux
Sanitary space	100 lux	100 lux	100 lux
Laundry	100 lux	100 lux	100 lux
Drying room	50 lux	100 lux	100 lux
Galley	300 lux *	200 lux	200 lux
Dry prov. store	100 lux	100 lux	100 lux
Passageways	50 lux	50 lux	100 lux

In the galley, the regulation of 300 lux is at working positions, while the designed illumination is referred as the minimum in that space. It is important to note that the required illumination is AVERAGE in the space, while the designed luminosity is referred as MINIMUM, and it is was tested during sea trials (see Figure 12). Moreover, there were various other spaces, without any illumination level requirements, that the ship-owner decided to specify, such as but not only, 300 lux in the wheelhouse and in the engine control room, 200 lux in the engine room workshop, etc.

Figure 12: Illumination level testing during sea trials.

The vibration and noise on board vessels are very critical in order to allow efficient workability and habitability. Exposure to noise can result in short-term and long-term hearing loss. Hearing loss can last from few minutes to a few days and can be also permanent and irreversible [6]. Frequent and regular exposure to vibration can result in short- and long-term problems such as pain, sleep disturbance, inability to do fine work, reduced grip strength and other problems [6].

The IMO criteria for noise [11] was optional till July 2014, when the new IMO Safety Of Life At Sea (SOLAS) amendment [12] was introduced to all vessels, therefore the subject PCTC vessel was designed meeting the latest rule requirements. The noise originated from the Main Engine and Propeller, in the accommodation is normally not an issue, because the accommodation is distanced from the engine room.

On the other hand, cargo ventilation fans are arranged around the superstructure, and air handling unit as well as refrigeration machineries are normally arranged inside the accommodation space, and these units introduce local noise which might interfere the social environmental condition. Cabins and spaces in the vicinity of machinery were designed with noise reduction wall panels, and the machinery spaces were designed with perforated noise reduction plates (see Figure 13). The noise all around the vessel was tested in the sea trial, and is tested in every subsequent sister vessel as well.

ISO vibration limit standard in merchant ships [13] was adopted in the TS stage, as a design target in the accommodation. During detailed design of the vessel, vibration analysis was carried out, and consequently, stiffening of high vibrating areas was introduced. A consideration to the vibration on board was the vibration induced from the propeller, and the effect of number of blades on the vibration. It was measured in model test and decided to go with the 5 blade propeller, having reduced pressure fluctuation to the stern body of the ship. The vibration test, similar to the noise test, was measured during sea trials, with mostly acceptable results in accordance with the agreed standard [13]. Stiffening to some extent was done when the vessel was alongside, based on finding local vibration higher than expected. The vibration test is carried out in every subsequent vessel in the series, considering that every vessel can behave differently.

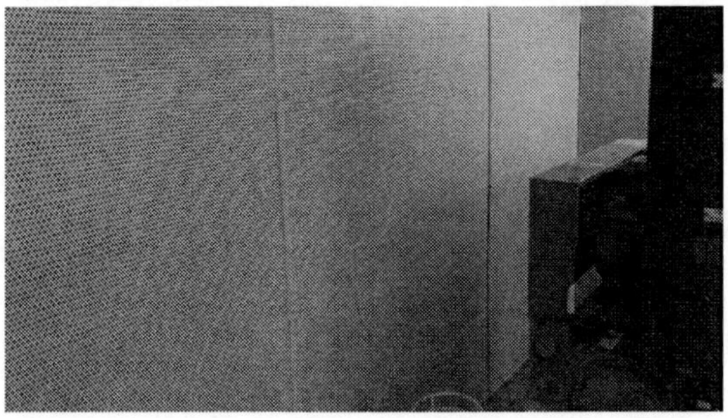

Figure 13: Perforated plate in the surrounding wall of the refrigeration machinery room, for reducing noise level in the vessel's accommodation.

Each cabin, for rating as well as for officers, was furnished with bed, wardrobe, desk with chair, settee with reception table, and private toilet (see Figure 14). The reason is that this owner philosophy is that happy crew will perform better, and improving the living space of the crew is a fundamental human element consideration.

Figure 14: Standard cabin. Note that curtains and upholstery are from fire retardant material.

Based on MLC 2006, a minimum of 7m^2 floor area should be designed for each single crew cabin, in ships above 10,000 Gross Tonnage (GT) [9]. The design of the subject PCTC vessel was such that the area of cabins of ratings was between 11.6m^2 and 12.1 m^2, with an average of 11.86 m^2. The area of cabins of junior officers was between 13.5m^2 and 17.5 m^2, with an average of 15.7m^2. The senior officers, chief engineer and the master cabins were designed with a dayroom and a bedroom for each person. It is therefore clear that in the case of a PCTC design, cabin's area is over the minimum requirement. This could be achieved without complications, because the accommodation in a PCTC vessel is normally on the upper deck, and has sufficient deck area to meet the accommodation design needs. Nevertheless, in order to improve the habitability on board, cabins were enlarged and more comfortable.

There are some other common spaces, which their areas were increased from the original design by the shipyard, based on owners understating and insisting about the improvement of habitability. An example is the Gym and Hobby room that was enlarged from 31.7m^2 to 35.3m^2.

The officers' nationality is mostly from Eastern European countries, while the ratings are mostly form the Philippines. In view of the diverse crew serving in the PCTC vessel, individualization was carried out. An example is the rice cooker added in the duty mess room, which is 24 hours opened, because the Philippine crew have the custom of consuming rice at any time of the day (see Figure 15).

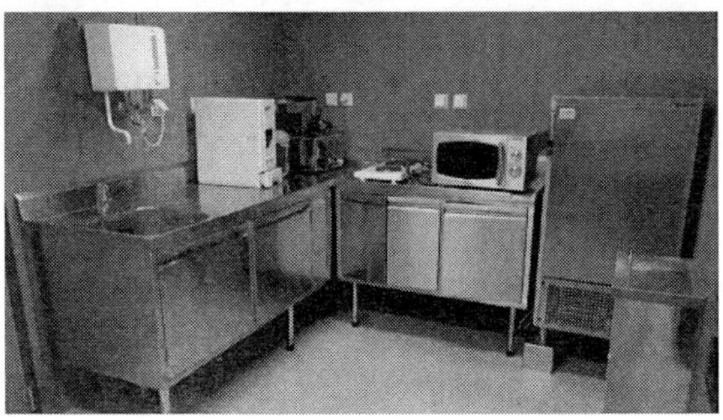

Figure 15: Duty mess is arranged with a rice cooker.

The elevator in PCTC vessels is usually arranged in the middle staircase, which goes from the engine room up to the accommodation deck. From the nature of this design, there is a need for a passageway in the upper deck from the lift to the accommodation entrance. An open-to-air walkway is used by many PCTC vessel owners. However, this is inconvenient and slippery in stormy weather, and therefore several PCTC vessel owners design a sheltered passageway with steel construction.

Figure 16: Transparent passageway from elevator to the accommodation lobby (left). Elevator push button in the lobby (right).

However, such a construction is subject to timely maintenance, to lack of natural illumination, and hot environmental condition. In the subject vessel, a transparent passageway was designed (see Figure 16) to counteract the problems arising from a steel structure, while keeping the crew dry and comfortable when walking between the accommodation and the elevator. An elevator calling push button was added in the accommodation lobby, to save waiting time for the elevator to reach the accommodation deck, as shown in figure 16.

4.3 MAINTAINABILITY

Creating changeable equipment was a critical consideration in the PCTC vessel design, and this philosophy applied for small spare parts as well as for large spares. In view of the natural environment that the vessel encounters, the material of equipment which is subject to corrosion, was designed from corrosion resistant materials, in order to be easily maintained. Therefore, the material of bolts, nuts, boxes (see Figure 17), etc., which are in touch with sea water or exposed to weather, was selected as marine type stainless steel (SUS316L), as long as the material has enough strength, whereas in other indoor locations these are treated with galvanization against corrosion.

Maintainability is examined also by the accessibility to every part of the vessel. Hence, in the engine room, being a confined space, during the detailed design stage of the vessel, it was made certain that each machinery was arranged with 360° access, that means that equipment is not attached to walls or bulkheads, but leaving enough space for crew member to access and maintain it.

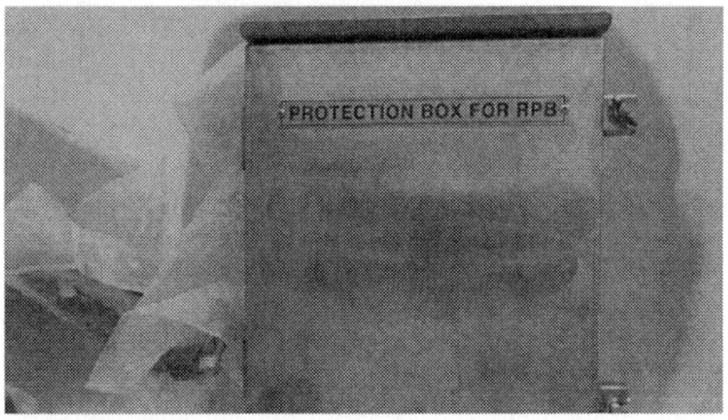

Figure 17: Marine type stainless steel box in weather deck.

On top of that, normally unattended spaces are designed by the yard not having sufficient accessibility. Nevertheless, in order to enhance the HF&E maintenance dimension, the ship-owner designer decided to provide the crew with safe and comfortable means to access such spaces. For example: vertical ladders were arranged inside each cargo hold ventilation trunk, with landing platforms between every couple of decks, and bolted type anti stowaway bars for enabling easy access from car deck to the trunk. Also, in the elevator trunk, vertical ladder and lighting fixtures were added (see Figure 18), for maintenance and also for safety escape route purposes. More so, the engine casing and funnel, gratings were fitted in every floor and all around each one of the exhaust pipes, in order to be fully and safely accessible by the crew.

Figure 18: Ladder and illumination in elevator trunk for safety and for maintenance. The view is from elevator top, downwards.

In the engine room, a large workshop was designed, and an ultrasonic cleaner was added to remove contamination from hard surfaces of machineries and general equipment (see Figure 19). Moreover, two ultrasonic soot breakers were added in the boiler for easy maintenance and cleaning of the exhaust gas tubes.

Figure 19: Ultrasonic cleaner in engine room workshop.

Separated spaces in the vicinity of the engine room workshop, were designed and constructed, in order to create an appropriate environment for maintenance, testing and working on specific tasks. For example: Electrical Workshop, Fuel Valve Test Room and Welding space.

Spare part transit in/out from the vessel to the engine room was considered in the design stage, in particular for large spare parts, such as thruster's motor, turbo charger, etc. As such, lifting equipment was design in accordance to the heaviest parts: Engine room crane was selected to lift the turbo charger as a complete unit out from the engine room, hatches were designed on the main deck (deck no.5) above engine room and bow thruster room, with enough size for the largest spare part to pass through, and lifting beams with electrical ratchet hoists were arranged in cargo deck no.5, in order to easily and safely take spares out from their rooms.

For light fixture replacement, the shipyard proposed in various locations hanging type lights, which are meant to be a solution against ship's vibration. However, the maintenance of hanging lights is more complicated because one seafarer needs to hold the fixture, and another to replace the bulb. Therefore, fixed type lights were insisted by the ship-owner designer, and the agreed solution against vibration was rubber shock absorbers in the mounting bolts (see Figure 20).

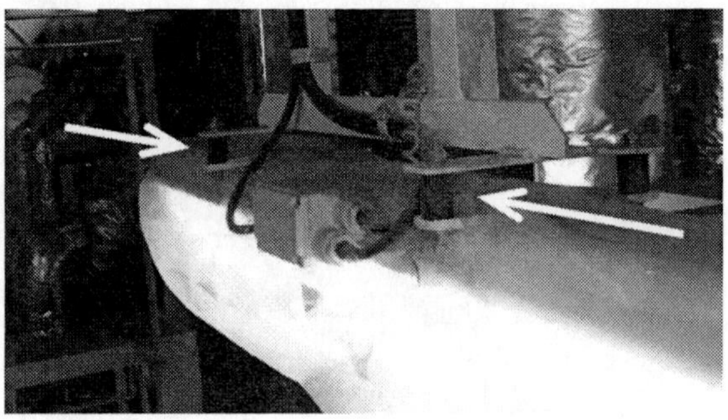

Figure 20: Lighting fixtures fitted with rubber shock absorbers against vibration.

4.4 SURVIVABILITY

The IMO regulates the safety of ships mainly though the SOLAS convention and amendments, however, the rules are considered as minimum standard and the ship-owner has the right to enhance the safety of the vessel and crew. In the subject PCTC design, various improvements were introduced, on top of the minimum requirements, based on the vast experience of operating a large PCTC fleet since 1993.

Firefighting wise, it was considered that fire, if start in accommodation, can initiate from inside a cabin, therefore fire detectors were added in every cabin even though detection is required only in corridors, as per rule requirement. Moreover, fire retardant curtains and upholstery was selected all over the accommodation, and fire hydrants with hoses were added inside the accommodation area.

The lifesaving equipment was designed in accordance with the International Life-saving Appliance (LSA) code, however, although one lifejacket per seafarer is a minimum requirement, the vessel was designed with two life jackets per person. The reason for that is that the crew's time is mostly divided between their cabins and the working spaces on board, therefore having one life-jacket in each cabin and additional one in the life-saving equipment store beside the muster station allows the crew to reach a life-jacket more immediately in an emergency situation.

On top of the minimum life jacket and immersion suits, lifesaving appliances for children were added, as well as jumbo size for a higher seafarer who might not fit in the standard size. Additional jackets and suits were prepared in the ship's bridge and engine control room, in order to give the crew the option for swift wearing of personal protection in emergency case.

Embarkation to the rescue boat in PCTC vessel can be very dangerous because the boat is on the upper deck, which is 31.023 meters above the lowest water line, thus a special embarkation procedure was designed from an outer shell door in cargo deck no.5, which means that the crew embark the rescue boat from 6.723 meters above the lowest water line, reducing dramatically the chances for accidents and injuries, as shown in Figure 21.

Figure 21: Rescue boat embarkation is designed from cargo deck number 5, close to the waterline.

The life boat is the PCTC vessel is a free fall type, which doesn't rely on a davit (see Figure 22) that might not work in emergency, and at the same time the free fall type has a quicker launching procedure.

Figure 22: Standard lifeboat arrangement on a PCTC – davit launched on upper deck.

For reducing slips in working decks such as mooring decks, steering gear deck, cargo deck (in some areas) and in exposed decks, they were all painted with anti-slip coating. For reducing cargo slips in cargo space the ramps and the loading/unloading paths, were coated with special rough anti slip coating, efficient against for heavy cargo slips and reducing possible accidents, see figure 8.

Handrails and storm rails were designed and arranged all around the superstructure, in engine room, and in staircases, were continues handrail with round transitions are fitted for the crew safety (see Figure 23). Moreover, vertical ladders above 2.5 meters were added with back support and protection bar. All these features were prepared to improve the crew safe traffic on board a moving vessel. For another safety feature see Figure 24.

Figure 23: Continues and smooth round shaped handrails in staircases, to prevent a case when the crew is leaving his grip, which may cause him to fall and be injured.

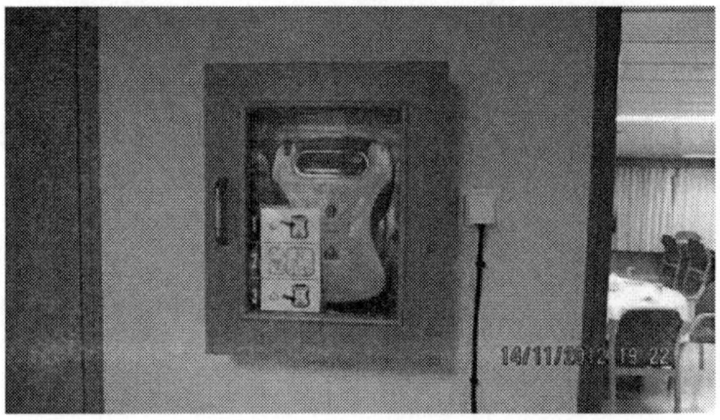

Figure 24: Defibrillator added in the vessel's accommodation as a lifesaving improvement for the seafarers.

4.5 CONTROLLABILITY

The controllability in the PCTC vessel's engine control room is considered by having one IMACS that is able to control and monitor the main machineries and systems in the vessel. This system is a user friendly one, with software that has been evolving in accordance to the feedbacks received from the crew. Its main idea is to avoid various control units for each piece of machinery, and integrate all into one place, controlled from the engine control room, from the bridge, and from the ship's office. Having this system on board the PCTC vessel, improved the ergonomics in terms of reach because there is no need to run around and have useless motions when all is done from one station (see Figure 25). The crew training also becomes easier and more efficient when one control station is arranged for all machineries.

Apart from the machineries control, another system is arranged for controlling the loading and stability of the case PCTC vessel and it's called LOADMATE. In lieu of hand calculations, which might be a burden to the chief officer, the stability is automatically calculated and controlled continuously from the loading PC, based on loading scheme inputted by the officer. It is interconnected to the vessel's draft and tank level gauge system and to the ballast control, in order to keep the vessel's stability in accordance to the SOLAS requirement. Moreover, the loading PC is controlling the anti-heeling system of the vessel, which is fitted to reduce heel angles and increase the safety of the vessel while loading/unloading cargo. Another advantage of this system, supplied by Totem Plus, is that the system can carry on a brief indicating inclining test before departing a port, in order to have a final confirmation about the vessel's condition, stability wise.

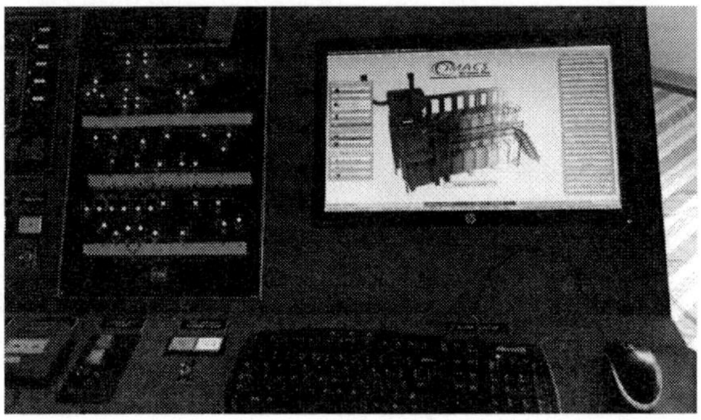

Figure 25: IMACS monitor in the bridge control console. Note that the keyboard, mouse and monitor are "of-the-shelf" pieces, which can be found in any port at a very low cost.

The control of the stern ramp was designed to be carried out from an enclosed room in the aft starboard side corner of the upper deck. The room was designed with windows all around, including wiper and heater, in order to allow the operator a smooth control over the stern ramp when lowered on the quay side and when hoisted back home.

In the PCTC vessel, having liftable decks, is it normal to have movable ramps inside cargo holds that can be adjusted to the different heights of decks. The control panels of every movable ramp were strictly designed to reduce the chances of accidents such as ramp being lowered on a crew and to allow continues control of the ramp during its manoeuvring. Hence, the controllers are arranged in such location that the operator can easily see the movement of the ramp (see Figure 26), and the ramp's main driver such as hydraulic piston, is also visible form the control panel. Visibility and reach were important factors when selecting the position and orientation of ramps control panels.

For enhancing the control on main spaces on board, a Closed Circuit Television (CCTV) system was designed and arranged in the vessel's bridge with cameras in mooring stations, in the entrances to the cargo hold from the quay side, in the engine room and in the pilot embarking area.

Figure 26: Movable ramp control panel oriented in such a way that the operator is facing the ramp for controlling the operation.

A key factor which enhances the controllability of the crew is standardization of equipment, in particular control equipment. Thus, the PCTC vessel bridge and engine control consoles were in essence designed similar to other PCTCs in the owner's fleet, in order to support familiarity of crew coming from other PCTC vessels in owner's fleet, while reducing the crew time of adaptation, avoiding misunderstandings and accidents.

4.6 AFFORDABILITY

It is essential that the design will be done correctly as much as possible from the start [1]. Therefore in the case of the PCTC vessel, initial design took most of the design time, and the TS was much detailed than originally prepared by the shipyard. The drawing approval stage also took huge effort from the ship-owner designer to implement all HF&E dimensions in the drawings. Nevertheless, as feedbacks keep coming back to the designer through the lifetime of the vessel, some modifications had to be applied retrospectively, and were costly.

Investment in high quality materials, in first class equipment suppliers, as well as automation of systems in the ship, is believed to reduce stresses from the crew, and improve their productivity. Sea water pipes were designed with polyethylene coating inside and lighting fixtures with LED type lights, are two examples to design specification with higher initial cost that pay back quickly because of the reduced maintenance and spare parts required during the operation of the vessel.

Moreover, the PCTC vessel design included investment in features that reduce the chances for accident or crew mistakes. Stern tube seal was designed with air chamber to prevent oil spill, rudder bush was designed with sea water lubrication instead of using oil/grease, for the same reason. More so, spare hydraulic oil storage tank for the bow thruster was arranged on top of the system itself, in order to be able to fill the system by gravity instantly, if needed, for example in case of a leak (see Figure 27).

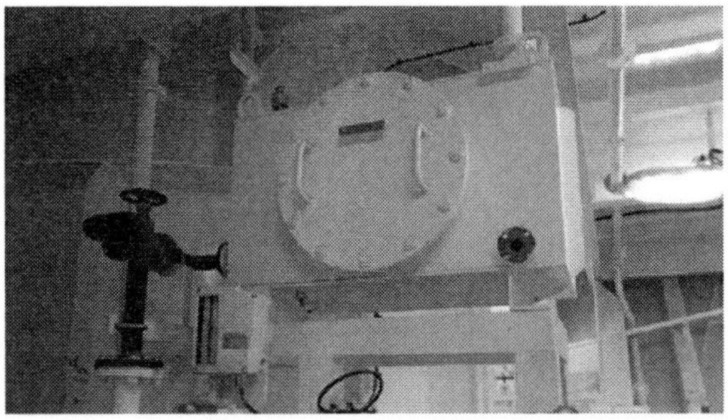

Figure 27: Spare hydraulic oil tank, for the bow thruster system.

4.7 COMMUNICATION AND USER EMPOWERMENT

Cultivating open communication between users and the design team enhances user empowerment and motivation [14].

This ship-owner has a sea trial fixed team, including experienced sea trial captain and chief engineer, who attend all new buildings sea trials. Both also take over every new delivered ship for about a month, so they get to see problems that could not been foreseen till the vessel started operating. On top of that, a designer from owner's office is always attending sea trial. The encounter of the designer with the end user always creates fruitful discussion with respect to possible improvements for the subsequent design.

From the other hand, once a vessel is delivered, frequent inspections throughout the vessel's lifespan are carried out by owner's inspection team, which improve the communication between the owner and the end user i.e. the crew. In these inspections the crew in many cases propose design improvements, which are transferred from the inspector to the designer for implementation.

Due to the above, the crew employed by this ship-owner is very much involved in the vessel's design, and when the crew sense that their comments are implemented, they feel pride to be part of such organization, and their motivation and dedication rise.

Additional examples of design improvements of the subject PCTC vessel initiated by the crew feedbacks are:

- In the accommodation the bonded store was switched with the stationary store, because the latter is better to be designed closer to the ship's office.
- Larger mess table to accommodate the crew comfortably.
- Change of cargo ramps sequence of opening/closing to allow better flow of cargo.
- Extend flag lines of the main mast for the vessel's flown flags to be more visible.
- Larger incinerator sludge tank, to reduce the frequent sludge burning.

- Engine control room was added with shelves to accommodate all manuals. More so, one of the windows was shifted to be have a more effective visibility to the engine room.
- Scupper pipes were added in the bridge wing, which apparently there were not enough for rainy condition.
- Etc.

4.8 PROFFESIONAL AND PSYCHOSOCIAL ASPECTS

Consideration was taken for better design to prevent human error, and protect the seafarer and the owner's reputation. Hence, the Electronic Chart Display and Information System (ECDIS) is arranged with a decision support tool to avoid collision (see Figure 28). It automatically analyzes the data of all targets and advises the officer on watch on the required action. The Closest Point of Approach (CPA) of all vessels is taken into account together with the stipulations of the International Regulations for Preventing Collisions at Sea, known as COLREGS. If required, a "Course to steer" advice will be given to the officer automatically in order to avoid collision [7].

A key point to reduce crew error is a good communication between all seafarers. Thus the internal communication system was improved much above the minimum requirements by rule, in way that telephones were added in almost every store, including: bosun store, mooring stations, near sounding pipes located inside cargo space, etc. Moreover, public address speakers were added in every car deck, for the crew to be well informed with the bridge announcements, when they are in the cargo space. The idea of adding internal communication means originates also from the shape of the vessel that makes the portable VHF in many areas useless.

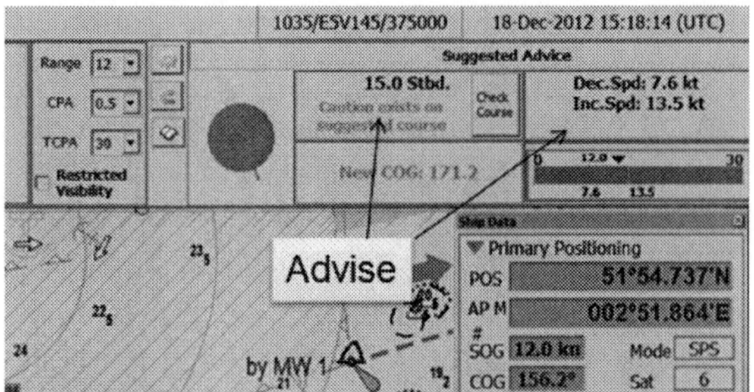

Figure 28: Screen shot from the decision support tool for collision avoidance. Note the advice to the officer to either change course or change speed, in order to avoid a collision.

The owner's design team also introduced various initiatives in the design of the vessel, aiming to increase the crew motivation and satisfaction, by having better working conditions, for instance:

- Swimming pool, sauna (see Figure 29) and high standard furnished gym and hobby room.
- Crew internet system, based of Wi-Fi, and supplied by Satmind, that can cost effectively communicate between the crew and their families while they are at sea.
- Elevator.
- Additional air conditioning units and air heaters in various stores and working spaces.
- Spacious accommodation cabins and public spaces.
- Etc.

Figure 28: Sauna for the crew.

5. SUMMARY AND CONCLUSIONS

This paper represent an example of how can a ship owner implement HF&E in a design of a new vessel. Various design improvements were presented with illustrations all for improving the ergonomic design of the case study vessel. It was noticed that the design imporvements are originated either from the crew based on their ideas on how to better the design, and also based on the ship owner design team based on their experience and innovations.

The most dominant design stage for improvements is the detailed design, mostly because of it being a prolonged stage where practical experience is relatively easy to apply. On the other hand, concept design and preliminary design, in the case study vessel were shorter, and concentrated on how to design a vessel to meet the market needs while keeping all current rule requirements.

A representing example is the design of main air compressors for a vessel in the three stages:

1. Concept design – Number of compressor and their capacity is mentioned, based on main engine requirement.
2. Preliminary design – Accessories are listed, cooling and lubricating type is selected.
3. Detailed design – Supplier is decided, full drawing of compressor are presented hence there is a large room for improving the design.

Based on the findings from this paper about participation of users in the design of a vessel, it is suggested that the ship owner designer will take to his tasks also coordination with the crew in order to obtain their feedback, and implement it in a new and possible retrofit projects.

It was also seen that as much as it is recommended to actively involve the users early in the design process [1], in practice the early design is sometime too short and mostly concentrated on market and regulations, rather than specific operational requirements from the crew. On the other hand, there are some occasions where the crew is involved at early stages, but to a lesser extent. One of the reasons for this phenomenon can be the perception that the crew is involved in day to day technical work, whereas the broad look of the shipping market and the need for vessels is on owner's shoulders.

Interestingly, at early design stages it seems more difficult to apply HF&E dimensions, because the main considerations are the shipping market and the regulations.

Other possibilities, to the participation of crew mostly in detailed design stage, are that today the new ships are very efficient, and therefore it is very hard to introduce changes during initial design stage. Moreover, at times of specialisation, it is believed that the crew are experts in vessel's operation, hence most of their experience can be more easily applied at detailed design stage, rather than initial design stages.

Directly involving the end user at any design stage of a new building is believe to be out of ordinary practice for shipyards and ship owners. Let alone the crew participation from early design stages – this is a major challenge for any ship owner.

The case study PCTC vessel design is a clear example on how HF&E can be implemented in a new building by a collaboration of the owner designers, the crew, and ship managers and the shipyard, and it is highly suggested that HF&E will be an integral part of the plan approval and design of every new vessel, in order to improve the various dimensions, which are Workability, Habitability, Maintainability, Survivability, Controllability, Affordability, Communication and user empowerment, and Professional and psychosocial aspects.

Disclaimer:
The opinions expressed in this paper are solely those of the author, and based on his experience.

6. REFERENCES

1. COSTA, N. A. & LUTZHOFT, M., *'The Values Of Ergonomics In Ship Design And Operation'*, RINA International Conference on Human Factors in Ship Design and Operation, London, 2014.

2. COSTA, N. A., *'Human Centred Design for Maritime Safety: A User Perspective on the Benefits and Success Factors of User Participation in the Design of Ships and Ship Systems'*, in Department of Shipping and Marine Technology, Chalmers University of Technology: Gothenburg, Sweden, 2016.

3. LLOYD'S REGISTER, *'The Human Element - An Introduction'*, Pamphlet, 2008.

4. RUMAWAS, V., *'Human Factors in Ship Design and Operation: Experiential Learning'*, in Faculty of Engineering, Science and Technology, Department of Marine Technology, Norwegian University of Science and Technology: Trondheim, Norway, 2016.

5. NOVAK, B., KIJORA, C., MALONE, T., LOCKETT-REYNOLDS, J. & WILSON, D., *'U.S. Department Of Homeland Security Human Systems Integration Applied To U.S. Coast Guard Surface Asset Acquisitions'*, International Conference on Human Performance at Sea, Glasgow, UK, 2010.

6. SINGH, R., *'The Ship and its* Environment', in Human Performance and Limitation for Mariners, The Nautical Institute, London, 2015.

7. TOTEM PLUS, *'http://www.totemplus.com/index.html'*, 2016.

8. THE BAHAMAS MARITIME AUTHORITY, *'Safe Manning Requirements'*, BMA Information Bulletin No.115, 2012.

9. ILO, *'Maritime Labour Convention'*, Geneva, 2006.

10. The Bahamas Maritime Authority, *'Merchant Shipping (Maritime Labour Convention) Regulation'*, 2012.

11. IMO, 'Code *on Noise Levels On Board Ships'*, IMO Resolution A.468 (XII), London, 1981.

12. IMO, *'Adoption Of The Code On Noise Levels On Board Ships'*, IMO Resolution MSC.337 (91) annex 1, London, 2012.

13. ISO, *'Mechanical vibration - Guidelines for the measurement, reporting and evaluation of vibration with regard to habitability on passenger and merchant ships'* ISO 6954, Geneva, 2000.

14. MAUNDER, A., MARSDEN, G., GRUIJTERS, D. & BLAKE, E., *'Designing Interactive Systems for the Developing World: Reflections*

on User-Centred Design', International Conference on Information and Communication Technologies and Development, Bangalore, India, 2007.

7. ACKNOWLEDGEMENTS

I wish to thank Ms. Nicole A. Costa, from the division of Maritime Human Factors and Navigation at Chalmers University of Technology, Sweden, for her contribution to me understanding of HF&E theories behind this paper.

8. AUTHORS BIOGRAPHY

Nicolas Bialystocki holds the current position of Project Manager at Antares Shipping Ltd. He is responsible for ships new buildings projects for Ray Car Carrier Ltd. including initial and concept design, technical specifications, makers selection, plans approval, inspections during construction and operation, sea trials of PCTC and Bulk Carriers built in Poland, Croatia, Vietnam and South Korea. Moreover, he is responsible for ship's innovations and research.

AUTHORS' CONTACT DETAILS

Anthony Woolley
Defence Science and Technology Group
anthony.woolley@dsto.defence.gov.au

Chantal Maïs
DCNS
chantal.mais@dcnsgroup.com

Chris Kelly
K Sharp Ltd.
chris.kelly@ksharp.co.uk

Christine Tomlinson
ABS
ctomlinson@eagle.org

Emil Dahle
Consultant
edahle@online.no

James Campbell
BMT Defence Services Ltd
JCampbell@bmtdsl.co.uk

Jean-Philippe Flohic
DCNS
jean-philippe.flohic@dcnsgroup.com

Jon Hill
Trident Marine
td@str.eu.com

Linda de Vries
Chalmers University of Technology
linda.de.vries@chalmers.se

Manjula Hettiarachchie
Colombo Dockyard PLC
manjula@cdl.lk

Margareta Lutzhoft
Australian Maritime College
Michelle.Grech@amsa.gov.au

Nicolas Bialystocki
Antares Shipping LTD
nicolas@antares.co.il

Simon Mockler
DNV GL
Simon.Mockler@dnvgl.com

Takeshi Shinoda
Kyushu University
shinoda@nams.kyushu-u.ac.jp

Trevor Dobbins
STResearch
td@str.eu.com